Orchids *of* Penang Hill

Rexy Prakash Chacko & Santhi Velayutham

THEHABITAT
PENANG HILL

This book was produced with the generous support of **Flagstaff Holdings Sdn Bhd, owner and operator of The Habitat Penang Hill.**

Published in Malaysia in 2020 by:
Entrepot Publishing Sdn Bhd
George Town, Penang, Malaysia
www.entrepotpublishing.com

Perpustakaan Negara Malaysia   Cataloguing-in-Publication Data

Rexy Prakash Chacko
Orchids of Penang Hill / Rexy Prakash Chacko & Santhi Velayutham.
ISBN 978-967-17008-5-3
1. Orchids--Malaysia--Bukit Bendera (Pulau Pinang)--Identification.
2. Orchids--Malaysia--Bukit Bendera (Pulau Pinang)--Classification.
I. Santhi Velayutham.
II. Title.
584.409595113

Design and layout by KODS Design

Printed by The Phoenix Press Sdn Bhd
6 Lebuh Gereja, George Town, 10200 Penang
Printed in Malaysia

Front cover: *Taeniophyllum hasseltii* Rchb.f.
Back cover: *Grammatophyllum speciosum* Blume

CONTENTS

MESSAGE FROM THE HABITAT

THE HABITAT PENANG HILL is delighted to be a sponsor of *Orchids of Penang Hill* and is honoured to work with the multi-talented authors, Mr. Rexy Prakash and Ms. Santhi Velayutham, and the amazing team at Entrepot Publishing, to bring this very important body of work to you. We trust that this labour of love will add to the collective knowledge of our natural world for the benefit of all.

The Habitat Penang Hill sits on the fringes of an ancient and untouched rainforest and was conceived from a deep love and respect for Mother Nature and for the Hill. We firmly believe that a physical connection to nature is central to one's understanding and appreciation of the natural world.

We are extremely blessed to have been given the opportunity to develop a world-class eco-tourism facility that all can be proud of, set in the midst of a virgin rainforest that is millions of years old and on a Hill steeped in history.

Hutan Simpan Kekal Bukit Kerajaan (Government Hill Permanent Forest Reserve) that surrounds our park was first gazetted in 1911 and is home to an estimated

2,500 species of flora and fauna. These include species that are endemic to Penang, such as the aptly named Penang Rock Gecko (*Cnemaspis affinis*). Indeed, this forest is a true treasure trove of biodiversity, as the results of our Penang Hill BioBlitz in October 2017 revealed. Regretfully, however, among the inhabitants of this rainforest, 20 are on the IUCN Red List, including four that are listed as Critically Endangered. This highlights the importance of our collective role as stewards of this precious green lung of Penang.

On 8 January 2016, The Habitat Penang Hill first opened its doors to the public with the premise of providing a space for visitors to contemplate, reconnect with and be amazed by the natural beauty of our tropical rainforest. Our hope is that through immersing themselves in nature, visitors will develop a sense of appreciation and respect for the forest, and through this process be inspired to make choices in their own lives for a more sustainable future.

© The Habitat Penang Hill

© The Habitat Penang Hill

The 1.6-km nature trail in our park was initially developed in the early 1800s by the British East India Company as a maintenance path for a drainage system designed to prevent landslides. Stepping inside the forest, you are immediately lulled by the cacophony of sounds; the rustle of leaves, the chirping of birds, and the buzz of insects all coming together in a mesmerising and deafening symphony.

Apart from the myriad flora and fauna that are discoverable within our forest, guests to our park are also able to enjoy breath-taking views of Penang and its surrounding areas from atop Curtis Crest Tree Top Walk. At 800 m above sea level it is the highest viewing point in Penang.

The Habitat is also home to Langur Way Canopy Walk, the longest two-span stressed-ribbon bridge in the world and the only one of its kind to be built within a pristine rainforest. With multiple viewing platforms, spanning approximately 230 m in length and standing at 15 m at the highest point, the bridge unlocks a unique view of the forest canopy, which is believed to host more than 50 per cent of the forest's biodiversity.

All excess funds generated by The Habitat park are channelled to support the work of our sister organisation, The Habitat Foundation, which is committed to supporting the conservation of biodiversity and safeguarding the living environment upon which we all depend. In addition, the Foundation helps to fund strategic conservation and research initiatives both within and beyond Penang and is working to establish Penang Hill as an important hub for rainforest and climate change research.

We work closely with communities, scientists, academic institutions, NGOs, government agencies, protected area managers, and businesses to achieve our objectives. Ultimately, our aim is to protect the irreplaceable natural heritage of this region and shift our society towards sustainability.

The Habitat Foundation complements The Habitat Penang Hill in fulfilling our objective of connecting people to nature and raising awareness of conservation through environmental education and community engagement.

We hope that you enjoy the Orchids of Penang Hill and look forward to welcoming you to The Habitat Penang Hill soon.

Sincerely,

Mr. Harry A. Cockrell
Co-Founder & Director
The Habitat Penang Hill
Co-Founder & Chairman
Board of Trustees
The Habitat Foundation

Mr. A. Reza Cockrell
Co-Founder & Director
The Habitat Penang Hill
Co-Founder & Chairman
Executive Committee
The Habitat Foundation

April 2020

FOREWORD

Penang Island is unique in that it is one of the most urbanised areas of Malaysia and yet it is also one of the most assessible places to nature. From any point in the island, you are within twenty minutes by car to a natural area; the sea or the forest. Penang Hill is the heart of Penang island's natural heritage. Its forest is rich in plant diversity and this book represents a dimension of this richness. While orchids are not particularly diverse in Penang island, their presence represents an important element of Penang's plant diversity. Although 144 species have so far been recorded on the island this is not a large number, as Peninsular Malaysia has close to a thousand species.

This book provides an illustrative account of fifty species of orchids found on Penang Hill. It gives a good selection. It also provides an excellent introduction to the orchids of Peninsular Malaysia. Some of them are particularly attractive and these are beautifully illustrated in the photographs found in the pages of this book. It will also serve as an easy and useful guide for visitors to Penang Hill.

Finally, congratulations to both the authors in their effort towards bringing this guidebook to publication.

Dr Saw Leng Guan
Curator
Penang Botanic Gardens

PREFACE

Penang Hill is the crowning jewel of Penang Island. A vital water catchment area, an ark of biodiversity and a refuge from the hustle and bustle of urban life, the hill, its charm and its cause have inspired people for generations. A key feature that makes Penang Hill so special is the diversity of flora and fauna which call this hill home; the hill's orchids with attractive and often unusual flowers epitomise this richness. Our inspiration to write this book began with gentle strolls along The Habitat Penang Hill's historically significant nature trail where we stumbled upon the beautiful and unusual orchids so different and diverse that were growing merely a few metres away from the trail. It immediately occurred to us that these orchids should be documented for better appreciation by the general public and, more importantly, as a key indicator of the richness that Penang Hill embodies.

The ambitious pursuit of documenting 50 selected orchids took us from the relative comfort of The Habitat Penang Hill's nature trail to the furthermost hinterland of the Penang Hill range. We scoured ridge and valley in search of these most enigmatic plants; to add to the complexity, not all orchids bloom for extended periods of time nor do they have set seasons. Thus, detailed planning, repeated visits and the luck of being in the right place at the right time were all needed to see and document each orchid in bloom. What we saw was nothing short of spectacular; the largest orchid in the world, *Grammatophyllum speciosum*, with thousands of flowers, *Dendrobium crumenatum* with hundreds of dove-like flowers filling the morning air with a sweet fragrance and the leafless *Taeniophyllum hasseltii* clinging on to small twigs, almost hidden from sight, all on the Penang Hill range.

We've ensured this book is illustrated with lots of colourful photographs to make it an easy-to-read guide, giving everyone an opportunity to behold and appreciate the sheer diversity of orchids on the hill. In this age where the climate crisis and unprecedented biodiversity loss are realities around us, we need to get to know the natural world better. Our hope is that this book will equip the reader with an understanding of our orchids and inspire them to be informed citizens who will rise to defend Penang Hill's biodiversity, away from the pursuit of unscrupulous development and short-term financial gain.

Rexy Prakash Chacko & Santhi Velayutham
March 2020

ACKNOWLEDGEMENTS

This book took shape through the efforts of many individuals and friends who helped us as we embarked on this ambitious pursuit to document the enigmatic orchids of Penang Hill. We thank God Almighty for giving us patience and tenacity throughout this project. Our greatest appreciation to The Habitat Penang Hill, especially the Cockrell family and Allen Tan, for supporting this project right from its inception. Their passion to see the conservation and preservation of Penang Hill, including its diverse orchids, is what inspired us to write on this subject and gave life to this book. Our utmost gratitude goes to Marcus Langdon and Keith Hockton of Entrepot Publishing, who most willingly agreed to publish this book. Remembering the late Kay Lyons, who meticulously proofed the book. Her contribution will always be cherished.

A special thanks to Ong Poh Teck of the Forest Research Institute Malaysia (FRIM), whose advice and expertise benefited us immensely while writing and reviewing this book. We record our appreciation to enthusiastic orchid spotters, Mohammed Mosharif and Lau Chai Thiam of The Habitat Penang Hill, who alerted us to blooming orchids in the wild which enabled them to be photographed and become cherished additions to the book. Our heartfelt thanks also to Ong Poh Teck, Tan Sin Hoong, The Habitat Penang Hill and The Tree Projects for their generous photo contributions to this book. Credit goes to Danesh Mapping Consultancy for the beautifully illustrated topographical map of Penang Island. Especial thanks to hiking companions Mike Gibby, Peter van der Lans, Rob Dickinson and Sim Boon Peng who accompanied us on many hiking expeditions in the wilderness of Penang, enabling us to spot and document many orchids for this book. Our sincere gratitude to our parents and families for their support and for nurturing this lifelong passion for orchids.

Lastly, this undertaking would have never been completed without the cooperation of all the orchids which bloomed at the right time and generously offered their beauty to colour the pages of this book.

INTRODUCTION

What are Orchids?

Comprising 10 per cent of all the world's flowering plants, the Orchid family is one of biggest and most charismatic members of the plant kingdom. Estimations of species count vary anywhere between 25,000 to 28,000 species, rivalled in number only by the Daisy family. The advent of orchid hybridisation in the nineteenth century revolutionised the orchid industry as laboratories and amateur hobbyists created thousands of hybrids.

Orchids are identifiable by their bilaterally symmetrical flowers, which consist of three sepals (sometimes the lateral sepals are fused into a synsepalum, e.g. *Paphiopedilum* and *Acriopsis*) and three petals. One of the petals is a highly modified one called the lip. In most orchids (apart from the primitive Apostasioideae subfamily), stamens and carpels are fused on a central column. The seeds are extremely small.

a) Parts of an orchid flower (Epidendroideae)

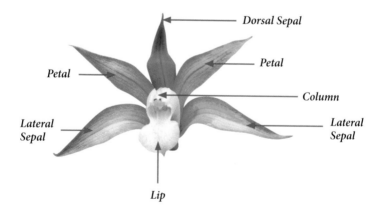

b) Parts of an orchid flower (Cypripedioideae)

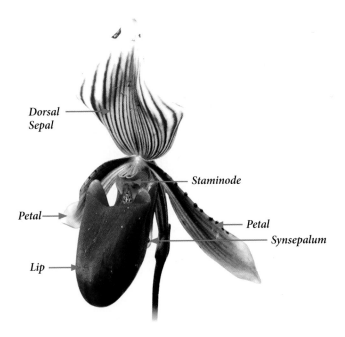

For centuries, orchids have been regarded as symbols of beauty, royalty and luxury. The term 'orchid' originates from the ancient Greek word 'orchis', which means 'testicle'.[2] It was coined by Greek philosopher Theophrastus, a disciple of Aristotle, in his book *History of Plants* due to the uncanny similarity of the shape of wild orchid tubers he found to that of a male testicle.

The ancient Chinese began cultivating orchids a few thousand years ago but orchids only became a popular houseplant in the early nineteenth century when specimens from the tropics first began appearing in England. It is believed that William John Swainson, a nineteenth-century explorer, unintentionally shipped some wild orchids from Brazil.[3] Shortly after arriving in England the plants bloomed, producing a dazzling flower display which triggered the Victorian-era orchid craze. The fame and glory of these plants were not confined to their exotic origins, but also their exquisite display of patterns, colours and shapes. This diversity was especially pronounced in the tropics, where adaptation to attract specific pollinators and to survive in niche habitats meant thousands of unique species existed.

Soon collectors were dispatched from England to the four corners of the world to collect orchids from the wild. They went into remote areas where disease, warring tribes and large carnivorous animals were widespread; as a result, many orchid hunters never arrived home alive. Nevertheless, thousands of orchids were collected by these hunters and shipped back to Europe, with few surviving the long and arduous journey. Frederick Sander, known as the 'Orchid King', had more than a dozen orchid hunters who scoured the globe to collect orchids for him. Rivalry between competing orchid hunters was rife, resulting in large numbers of wild plants being indiscriminately destroyed to prevent another party collecting them.

The unsustainable and indiscriminate collection of orchids pushed many species to the verge of extinction. The discovery of seed germination techniques slowly relieved the pressure on wild populations, but today poaching and habitat loss remain existential threats to many wild orchid populations around the world.

Classification of Orchids

The Orchid family, Orchidaceae, falls taxonomically under the class Monocotyledonae within the superorder Liliiflorae. Under this umbrella, Orchidaceae is placed in the Asparagales order, where it forms a monophyletic group (descendants of the same ancestor) that consists of five subfamilies.

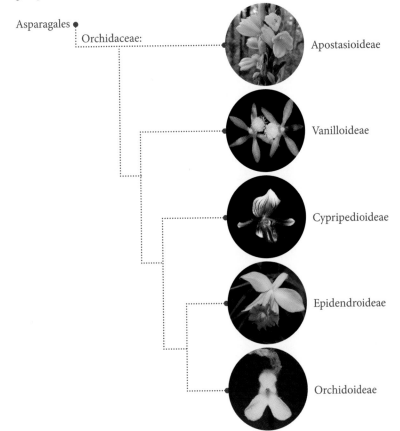

Evolutionary relationships among the five subfamilies of Orchidaceae.

Apostasioideae

This subfamily is the most primitive. Flowers are lily-like and have free stamens with two or three fertile anthers, pollen in loose powdery form, not forming pollinia.

Vanilloideae

This subfamily has a single fertile anther, loose pollen, sometimes tetrads or rarely forming true pollinia without accessory structures. The most famous member of this subfamily is *Vanilla planifolia*, the source of natural vanilla. Other members of this subfamily include the holomycotrophic (plants that do not have chlorophyll but obtain nutrition from fungi) genera such as *Cyrtosia, Erythrorchis, Galeola* and *Lecanorchis*.

Cypripedioideae

This subfamily, commonly known as the Slipper Orchid, consists of species characterised by flowers with pouch-like lips, column with two fertile stamens, pollen usually paste-like, powdery or viscous, staminode conspicuous and shield-shaped.

Epidendroideae

The largest subfamily of orchids, consisting of species with a single fertile anther, pollen coherent forming definite pollinia.

Orchidoideae

This subfamily consists of terrestrial or, rarely, epiphytic plants with root tubers, stem tubers or fleshy, short to long rhizomes. Some are holomycotrophs without (or with much reduced) leaves. The pollen is coherent and forms definite pollinia.

This book describes 50 species of orchids which are native to Penang Hill. They are arranged according to the order of these five subfamilies, within which the genera and species are arranged in alphabetical order.

What's in an Orchid Name?

Orchid naming follows the binomial nomenclature, Carl Linnaeus's two-term naming system for all living things. For example, the Pigeon Orchid, *Dendrobium crumenatum*. *Dendrobium crumenatum* is the scientific name, which follows the binomial nomenclature, where the first term, *Dendrobium*, is the **genus** of the orchid, while the second term, *crumenatum*, is the **species** of the orchid, its specific identity. Most orchids are easily identifiable to their genus by the leaves, stems or pseudobulbs, but identifying the species often requires the plant to be flowering. Orchids which are culturally and medicinally significant usually have a common name. *Dendrobium crumenatum* is commonly known as the Pigeon Orchid.

However, not all orchid species in the wild have common names. Common names of the same plant can differ from region to region; likewise, the same common name can be used to describe more than one species. Thus, scientific names have precedence over common names as they can dispel confusion about a plant's identity. Scientific names of the 50 species in this book are based on the January 2020 inventory of names obtained from the *World Checklist of Selected Plant Families* (WCSP).[5]

Dendrobium (genus) *crumenatum* (species), the Pigeon Orchid (common name).

Habit

Orchids are broadly categorised by three lifestyle choices based on the niche space in which they grow – terrestrials, epiphytes and lithophytes (or in a combination of any or all three categories).

Terrestrial Orchids

Terrestrial orchids grow on the ground. Just like most other ground-dwelling plants, they derive their nutrients and water from the soil. While most orchids photosynthesise, a small number of terrestrial orchids do not. Collectively known as holomycotrophic orchids, these draw all their nutrients from fungi.

Eulophia nuda, growing terrestrially. >

Epiphytic Orchids

Epiphytic orchids grow on the surface of other plants, e.g. tree trunks, shrubs and even leaves. They do not impact their host negatively as they only use the host for physical support; they derive nutrients and water from the surroundings. A majority of orchids in the tropics are epiphytic.

Grammatophyllum speciosum, growing > epiphytically on a large tree.

Lithophytic Orchids

Lithophytic (or epilithic) orchids grow on rocks. Many grow in dry and sunny exposed conditions, making them naturally hardy. They derive nutrients and water from rain and decaying material on the rock surface. They face many similar challenges to epiphytic orchids and thus quite a number of species occur naturally as both epiphytes and lithophytes.

Bulbophyllum gracillimum, growing as a > lithophyte on a granite boulder.

How to use this guide?

The following pages of this guide highlight a representative selection of 50 orchid species that are found on Penang Hill. The species are arranged by order of the five Orchidaceae subfamilies, within which the genera and species are arranged in alphabetical order. Colour photographs and summaries of key attributes, adapted mainly from *The Orchids of Peninsular Malaysia and Singapore* (1992) by Gunnar Seidenfaden and J.J. Wood, give the reader a better understanding of each species. The following is an explanation of the format used in this book.

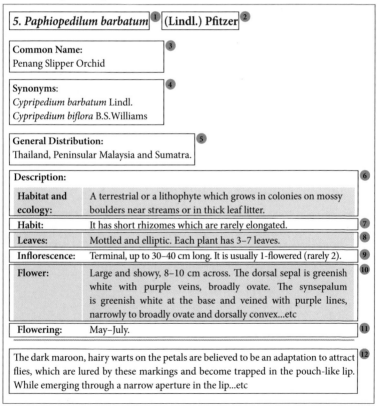

5. Paphiopedilum barbatum (1) (Lindl.) Pfitzer (2)

Common Name: (3)
Penang Slipper Orchid

Synonyms: (4)
Cypripedium barbatum Lindl.
Cypripedium biflora B.S.Williams

General Distribution: (5)
Thailand, Peninsular Malaysia and Sumatra.

Description: (6)	
Habitat and ecology:	A terrestrial or a lithophyte which grows in colonies on mossy boulders near streams or in thick leaf litter.
Habit:	It has short rhizomes which are rarely elongated. (7)
Leaves:	Mottled and elliptic. Each plant has 3–7 leaves. (8)
Inflorescence:	Terminal, up to 30–40 cm long. It is usually 1-flowered (rarely 2). (9)
Flower:	Large and showy, 8–10 cm across. The dorsal sepal is greenish white with purple veins, broadly ovate. The synsepalum is greenish white at the base and veined with purple lines, narrowly to broadly ovate and dorsally convex...etc (10)
Flowering:	May–July. (11)

The dark maroon, hairy warts on the petals are believed to be an adaptation to attract (12) flies, which are lured by these markings and become trapped in the pouch-like lip. While emerging through a narrow aperture in the lip...etc

1. **Scientific name.** Follows the binomial nomenclature consisting of the name of the genus (first term) and the species (second term).

2. **Author citation.** Refers to the individual(s) who validly published the botanical name.

3. **Common name.** The colloquial or local name of the plant. It may be in English or Malay.

4. **Synonym.** A name previously used as the correct scientific name but has been displaced by another name. However, not all plants have synonyms.

5. **General distribution.** A geographical area where the species can be found. However, this list may not be exhaustive.

6. **Habitat and ecology.** Categorises the plant as either an epiphyte, a lithophyte or a terrestrial (or in a combination) and gives information on the likely setting of the plant's habitat.

7. **Habit.** Describes the plant's stems, rhizomes or pseudobulbs.

8. **Leaves.** Describes the nature of the plant's leaves.

9. **Inflorescence.** Describes the flower head of the plant.

10. **Flower.** Describes the flower, including details of its colour and shape.

11. **Flowering.** Describes the flowering season of the plant (based on observations in Penang). However, some species do not have a set flowering season.

12. **Trivia.** Interesting information on cultural, medicinal and other uses or facts. However, not all species include a trivia section.

The Orchid Code

It is important that a reader understands his or her responsibility when using this guide. While some might enjoy flipping through its pages as a pastime, others might be more serious in wanting to spot orchids in the wild. Spotting orchids in the wild is extremely satisfying, especially when one is lucky enough to see them in flower. However, while orchid spotting is fun, it is important that it is done responsibly with the right intentions. 'The Orchid Code' lists several tips and best practices one should adhere to while using this guide.

- Use this guide to spot and recognise orchids in the wild. Orchids are best appreciated in their natural habitat. If you are lucky enough to have them growing naturally in your property, conserve the immediate surroundings of the plant so that it can thrive.

- When spotting orchids in the wild it is important not to disturb the plant and its surroundings or indiscriminately trample on vegetation nearby in order to get a better view or a photo. Disturbance of the immediate surroundings results in the alteration of the microenvironment in which the plant thrives. Many orchids exist as colonies with smaller plants around the mother plant, and thus exercising caution is a must in order not to trample and damage young plants.

- Do not use this book as a guide to poach wild orchids. Orchid poaching is a crime. It is a wasteful exercise because a collected specimen, without the right conditions, will not flower, and quite often will die of shock and heatstroke in the new surroundings. Under the National Forestry Act of 1984, orchid poaching in protected forests is illegal.

- When you come across wild orchid colonies, do not reveal the specific location publicly. Such revelation could be deadly to the plant as the information could be used by collectors to decimate wild populations.

- The medicinal information written in this book is compiled from various sources; many are traditional remedies. However, they may not have been scientifically tested for their safety and efficacy. Thus, do not attempt to use them without a medical doctor's approval. If uncertain, seek help from those with more knowledge. The authors cannot be held responsible for any untoward incidents arising from the indiscriminate use of the information in this book.

- Always seek to learn more about the orchids that you find. Referring to an orchidologist or a more comprehensive orchid book will help improve understanding and appreciation of the plant.

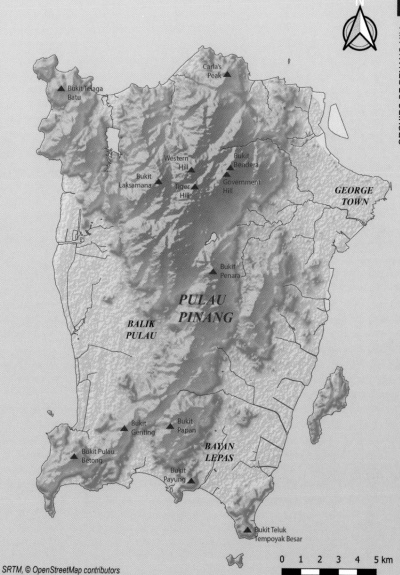

N

Carla's
Peak

Bukit Telaga
Batu

Western
Hill

Bukit
Bendera

Bukit
Laksamana

Tiger
Hill

Government
Hill

GEORGE
TOWN

Bukit
Penara

PULAU
PINANG

BALIK
PULAU

Bukit
Genting

Bukit
Papan

BAYAN
LEPAS

Bukit Pulau
Betong

Bukit
Payung

Bukit Teluk
Tempoyak Besar

SRTM, © OpenStreetMap contributors

0   1   2   3   4   5 km

PENANG HILL – THE GREEN HEART OF THE ISLAND

The island of Penang on the north-western coast of Peninsular Malaysia has a hilly topography. In fact, 50 per cent of the land on the island is hilly, underlain mainly by a granitic bedrock which forms the Penang Hill range. This central ridge of hills runs from the south-western tip to the northernmost point of the island. The hill range steadily rises in elevation from modest 200–300 m above sea level (asl) peaks in the southern tail of the range near Gertak Sanggul to 700–800 m asl peaks in the Penang Hill summit region, reaching its highest elevation at 833 m asl at the peak of Western Hill. The northern region of the range, popularly known as Penang Hill, is a series of narrow, branched ridges with peaks above 700 m asl such as Bukit Laksamana, Tiger Hill, Western Hill, Government Hill and Bukit Bendera.

The forests of Penang Hill, a biodiversity hotspot. © The Habitat Penang Hill.

Land use in the entire range varies, with a large portion of the southern and central regions occupied by vegetable farms and fruit orchards with vestiges of primary forest clinging to isolated pockets on ridge tops and peaks. The northern portion, where Penang Hill is located, is still largely hill dipterocarp forests with a small strip of lower montane forest at the summit area.[6] These forests are protected as a network of forest reserves and water catchment areas, which has guaranteed that the original vegetation is preserved.

The difference in land use, as well as the higher elevation of Penang Hill, has ensured that this region receives a substantially higher amount of rain and slightly milder temperatures than the lowlands and the rest of the range. These features correspond with a markedly higher abundance in orchid species that still exist in the Penang Hill area compared to the rest of the range. Thus, the continued protection and conservation of Penang Hill and its natural ecosystem are pivotal in ensuring the survival of orchid species on the Hill.

The government and civil society must play an important role in ensuring the continued conservation of Penang Hill for generations to come. The proposed gazettement of Penang Hill as a UNESCO Biosphere Reserve is a step in the right direction, ensuring not only the orchids, but all flora and fauna as well as the precious ecosystem services that Penang Hill provides, are guarded for future generations.

Hill dipterocarp forest on Penang Hill.
© The Habitat Penang Hill.

Orchids of Penang Hill – A Background

Soon after Francis Light set foot in Penang, paths to the summit area of Penang Hill were established. These paths enabled access into once impenetrable jungle and facilitated pioneering botanical studies from the 1790s. Many plants from Penang Hill were collected, catalogued and shipped to herbariums around the world.[7] Early botanists who conducted surveys in Penang, like Nathaniel Wallich, George Porter and Alexander Maingay, all reported orchids in their records from Penang Hill. The most notable of these early botanists in Penang was Charles Curtis who was appointed the first Superintendent of the Penang Botanical Gardens. His pioneering 1894 book entitled *A Catalogue of the Flowering Plants and Ferns Found Growing Wild in the Island of Penang* listed 1971 species, of which 90 species were orchids.[8] H.N. Ridley, another renowned botanist, contributed significantly to the identification of orchids from Penang Hill. More than a century after the publication of Curtis' book, in 2011 an article by Rusea Go and eight other authors titled 'An assessment of Orchids' Diversity in Penang Hill, Penang after 115 years', was published after an assessment of orchid diversity on Penang Hill from 2004 to 2008.[9] In this study the cumulative tally of orchids recorded from Penang Hill was 136 species.

The work by Rusea Go and her fellow researchers was continued by researchers Farah Alia Nordin and Nga Shi Yeu. Their work brought the tally in 2017 to 144 species.[10] However, the real figure may only be known through a more comprehensive study. Only one endemic orchid is known to occur in Penang Hill, *Zeuxine rupestris*. It is a species described by H.N. Ridley in 1903.[11] The showy *Paphiopedilum barbatum*, probably the most enigmatic orchid in Penang, was once widespread on the slopes of Penang Hill, earning it the common name 'Penang Slipper Orchid'. Species named after 'Penang' such as *Ania penangiana* and *Polystachya penangensis* (now a synonym of *Polystachya concreta*) also point to the fact that the type locality of these orchids would have been Penang, quite possibly Penang Hill.

Orchids of Penang Hill

Our Orchids are in Peril!

The beauty and allure of orchids have fascinated mankind from time immemorial. Quite ironically, it is this very beauty that has brought peril to them. Collecting from the wild is a direct threat, be it by amateur collectors, traders or private nurseries. The indiscriminate collecting of orchids by orchid hunters in the nineteenth and early twentieth centuries decimated wild populations, especially the showier and exotic ones. Penang also had a role in the wild orchid trade; the 1891 edition of the Straits Settlements Annual Report noted that the Penang Botanic Garden's principal excursion to Langkawi 1890 brought back 'several thousand specimens' of *Paphiopedilum niveum*, a disproportionately large number for a species which has a relatively restricted geographic distribution.[12]

Frederick Sander's firm, which dispatched orchid hunters around the world, also came to Penang and collected a huge *Grammatophyllum speciosum* which weighed one ton.[13] This and many other orchid collecting excursions depleted wild populations, most notably the *Paphiopedilum barbatum* (Penang Slipper Orchid). An article by K. Sim entitled 'Flowers of Penang Hill' which appeared in the *Singapore Free Press* in 1950 described how one can come across 'occasionally the sumptuous maroon and purple slipper orchid' (a reference to *Paphiopedilum barbatum*) along the jungle paths on Penang Hill.[14] However, today this species is extremely rare on the trails which meander Penang Hill, tethering at the brink of extinction because of unscrupulous collection.

Some private orchid nurseries are notorious for hiring local hunters to strip wild populations and then proceed to sell these wild specimens openly. With the advent of social media, illegal orchid harvesting and trade has expanded, with collectors – both private nurseries and amateur hobbyists – using social media platforms as a means of showing off their poached specimens and reaching out to potential buyers. This online trade has been relatively unregulated as it is difficult to monitor. What is more worrying is that such online postings influence more people to go collecting, further depleting wild populations.

Rampant buying and selling of wild orchids via social media platforms.

Habitat loss remains a paramount danger to wild orchids as activities such as logging, quarrying, agriculture and residential development uproot wild orchids, disrupt the microclimate and rob them of the ecosystem which they need in order to grow.

Forest clearing destroys orchid habitat and alters the microclimate orchids need to survive.

In Penang, quarrying, rampant agriculture (both legal and illegal) on the Penang Hill range as well as creeping residential development in hilly areas are key threats to the survival of wild orchids. Rusea Go notes in her 2011 article that many of the localities where Charles Curtis recorded orchids in his 1894 publication have since been converted to residential and agricultural land.[15] This possibly explains the absence in the recent study of repeat collections of certain species recorded by Charles Curtis.

Save Our Orchids!

Urgent conservation and preservation measures are needed to ensure orchids do not head down the path of extinction. Better legislation to protect orchid species and habitat is needed as well as more stringent monitoring of the online orchid trade. As members of the public and responsible consumers, we should not collect or buy poached wild orchids. They can be recognised from their irregular shape and size, broken and disorganised roots as well as leaves covered with lichens.[16] Giving a helping hand to the authorities by reporting cases of wild orchid poaching to the State Forestry Department is important, too, as it gives vital information to stop the illegal orchid trade. All of us have a role to play in ensuring the survival of orchids because when we lose an orchid, we not only lose a beautiful flower but, more critically, we lose an important member in the intricate web of life.

APOSTASIOIDEAE

1. *Apostasia nuda* R.Br.

Common Name:
Kencing Pelanduk

Synonyms:
Adactylus nudus (R.Br.) Rolfe
Apostasia brunonis Griff.

General Distribution:
India, Bangladesh, Myanmar,
Vietnam, Thailand, Peninsular
Malaysia, Sumatra and Borneo.

Description:

Habitat and ecology:	A terrestrial distributed from lowland to hilly areas which generally grows in the undergrowth.
Habit:	The stem is erect, 30–50 cm long.
Leaves:	Numerous, narrow, smooth and spirally arranged. Often varying in size.
Inflorescence:	Terminal, sometimes lateral and branched. It emerges horizontally and becomes pendulous, borne on the axils of shortened upper leaves, with 20–75 flowers.
Flower:	Wide opening, entirely white (or yellow). The sepals and petals are narrowly elliptic and rolled backwards. The lip is narrowly elliptic while the column is bent at an angle.[17]
Flowering:	Throughout the year.

In Peninsular Malaysia, the boiled roots are used as a poultice to treat diarrhoea, and also applied to treat dog bites.[18]

2. *Apostasia wallichii* R.Br.

Common Name:
Yellow Grass Orchid

Synonyms:
Apostasia curvata J.J.Sm.
Mesodactylis wallichii (R.Br.) Endl.

General Distribution:
Thailand, Peninsular Malaysia,
Sumatra, Java, Borneo, New Guinea
and Queensland.

Description:

Habitat and ecology:	A terrestrial which grows on shaded slopes in hilly areas.
Habit:	The stem is erect and slender, growing 20–40 cm long.
Leaves:	Closely set, smooth and narrow.
Inflorescence:	Terminal, 5–10 cm long. It emerges from the upper leaf axil and is erect or pendulous, bearing 20–100 flowers.
Flower:	Wide opening, entirely bright yellow and star-shaped. The sepals and petals are narrowly elliptic. The lip is narrowly elliptic, the column is straight with a staminode that is not winged at the base.[19]
Flowering:	April–August.

Traditionally the leaves are used to treat diabetes, relieve fevers, mouth sores, skin diseases and asthma.[20]

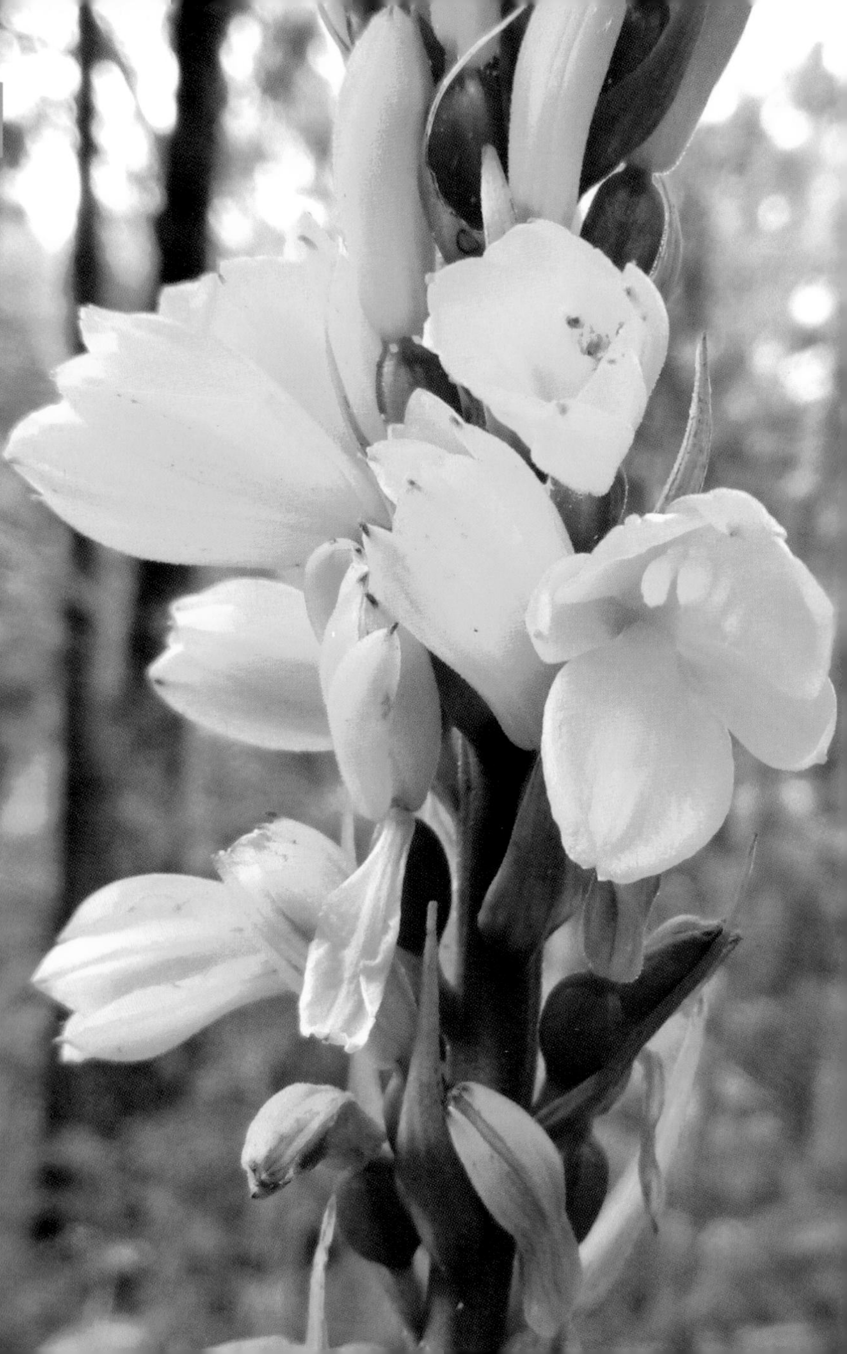

3. *Neuwiedia veratrifolia* Blume

Common Name:
Common Neuwiedia

Synonyms:
Neuwiedia cucullata J.J.Sm.
Neuwiedia lindleyi Rolfe

General Distribution:
Peninsular Malaysia, Sumatra, Java,
Borneo, Maluku, New Guinea
and Vanuatu.

Description:

Habitat and ecology:	A terrestrial which is commonly found in exposed secondary growth in hilly areas.
Habit:	It has a stout rhizome up to 20 cm long.
Leaves:	Narrowly elliptical and palm-like.
Inflorescence:	Terminal, growing to more than 50 cm long. It is stiffly erect and unbranched, holding up to 60 flowers.
Flower:	Not opening widely, fragrant, entirely bright yellow. The sepals are oblong-ovate and slightly hairy on the outside. The petals are oblanceolate and glabrous. The oblanceolate lip has a raised longitudinal midrib.
Flowering:	May–October.

VANILLOIDEAE

4. *Lecanorchis multiflora* J.J.Sm.

General Distribution:
China, Thailand, Peninsular Malaysia, Sumatra, Java and Borneo.

© Ong Poh Teck

Description:

Habitat and ecology:	A terrestrial holomycotroph which grows in leaf litter from lowland to hilly areas.
Habit:	The plant has slender, maroon stems up to 60 cm long.
Leaves:	It has no leaves and does not photosynthesise, instead relying on nutrients from its fungal associates to survive.
Inflorescence:	Terminal, growing to 6 cm or more. It is erect, producing 5 or more flowers which are closely arranged together in succession.
Flower:	Wide opening, upward facing. The sepals are yellowish green, narrowly oblong-obovate. The petals too are yellowish green, oblong-elliptic. The tip of the trilobed lip is covered in a dense mass of white hairs.

CYPRIPEDIOIDEAE

5. *Paphiopedilum barbatum* (Lindl.) Pfitzer

Common Name:
Penang Slipper Orchid

Synonyms:
Cypripedium barbatum Lindl.
Cypripedium biflora B.S.Williams

General Distribution:
Thailand, Peninsular Malaysia
and Sumatra.

Description:

Habitat and ecology:	A terrestrial or a lithophyte which grows in colonies on mossy boulders near streams or in thick leaf litter.
Habit:	It has short rhizomes which are rarely elongated.
Leaves:	Mottled and elliptic. Each plant has 3–7 leaves.
Inflorescence:	Terminal, up to 30–40 cm long. It is usually 1-flowered (rarely 2).
Flower:	Large and showy, 8–10 cm across. The dorsal sepal is greenish white with purple veins, broadly ovate. The synsepalum is greenish white at the base and veined with purple lines, narrowly to broadly ovate and dorsally convex. The petals are purple with dark maroon hairy warts on the upper margins, linear-oblong, slightly wider near the apex. The prominent purple-maroon lip is pouch-shaped. The staminode is green suffused with purple, crescent-shaped and sparsely hairy.
Flowering:	May–July.

The dark maroon, hairy warts on the petals are believed to be an adaptation to attract flies, which are lured by these markings and become trapped in the pouch-like lip. While emerging through a narrow aperture in the lip, the fly gets smeared in pollen. When that fly visits another flower, the process is repeated, and the flower gets pollinated. Once common near streams, this species is on the brink of extinction on Penang Hill. Globally, it is listed as endangered on the IUCN Red List of Threatened Species.

EPIDENDROIDEAE

6. *Acampe praemorsa* var. *longepedunculata* (Trimen) Govaerts

Common Name:
Stiff Acampe

Synonyms:
Acampe rigida (Buch.-Ham. ex Sm.) P.F.Hunt
Acampe penangiana Ridl.

General Distribution:
China, Taiwan, Vietnam, Laos, Myanmar, Thailand, Peninsular Malaysia and the Philippines.

Description:

Habitat and ecology:	An epiphyte or a lithophyte widely distributed from lowland to hilly areas.
Habit:	It has stout stems which are often branched.
Leaves:	Stiffly ascending, thick with broad tips.
Inflorescence:	Arises between axils of the leaves. It is erect with short side branches bearing 8–10 closely-arranged flowers.
Flower:	Not opening widely, 2–3 cm across, and upward facing. The fleshy sepals and petals are yellow with narrow horizontal red stripes. The dorsal sepal is broadly rounded at the end. The lateral sepals are almost the same but keeled. Petals are narrower than the lateral sepals. The upward-pointing lip is white and hairy on its inner side. The short column has small horns.
Flowering:	September–October.

In 2012, the Chinese Academy of Sciences discovered that the plant can employ rain to self-pollinate. Raindrops flick away the anther, causing the pollen to be ejected upward and finally fall into the stigmatic cavity, pollinating the flower.[21] Laotians use the leaves to make mats.[22]

7. *Acriopsis indica* C.Wright

General Distribution:
India, Myanmar, Thailand, Peninsular Malaysia, Borneo and the Philippines.

Description:

Habitat and ecology:	An epiphyte or a lithophyte which usually grows in clumps on tree branches near the canopy.
Habit:	It has closely set, ovoid and ridged pseudobulbs which usually form dense clumps. The bulbs are 2.5–5 cm long.
Leaves:	Narrowly oblong, 5–9 cm long. Each bulb has 2–3 leaves.
Inflorescence:	Arises from the basal node of the pseudobulb, growing to 30 cm long. It is erect, branched and holds many well-spaced flowers.
Flower:	Wide opening, about 1 cm across. Sepals and petals are greenish-yellow with red blotches. The dorsal sepal is narrow. The lateral sepals are joined into a synsepalum. The petals are broader than the dorsal sepal. The lip is white with purple blotches and has small wavy edges.

8. *Aerides odorata* Lour.

Common Name:
Fragrant Cat's Tail Orchid

Synonyms:
Aerides ballantiniana Rchb.f.
Aerides cornuta Roxb.

General Distribution:
India, China, Vietnam, Thailand, Peninsular Malaysia, Sumatra, Borneo and the Philippines.

Description:

Habitat and ecology:	An epiphyte or a lithophyte distributed from lowland to hilly areas.
Habit:	Its stout, long stems droop downwards.
Leaves:	Fleshy and curved, 25–30 cm long.
Inflorescence:	Originates from the leaf axils and can grow 25–35 cm long. It is pendulous and supports 15–30 flowers.
Flower:	Wide opening, about 3 cm across, fragrant. The sepals and petals are white with violet-purple blotches at the tips. The dorsal sepal is elliptic. Lateral sepals are broadly ovate. The petals are elliptic. The lip has a spur shaped like a horn which is mildly green or yellow with a prominent purple midline. The tip of the spur contains nectar.
Flowering:	Throughout the year.

The sweet fragrance of this species and its easy-to-grow nature make it popular with orchid enthusiasts. Ground and mixed with neem tree bark paste, the roots are used to cure joint pain and swelling. [23]

9. *Agrostophyllum majus* Hook.f.

Synonyms:
Agrostophyllum denbergeri J.J.Sm.
Agrostophyllum agusanense Ormerod

General Distribution:
Thailand, Peninsular Malaysia,
Sumatra, Borneo, Solomon Islands
and Vanuatu.

Description:

Habitat and ecology:	An epiphyte or a lithophyte commonly found on exposed trunks and branches of trees in hilly areas.
Habit:	The stems are long and flattened, up to 100 cm long.
Leaves:	Obtuse, arranged at acute angles to the stem and grow 20 cm long.
Inflorescence:	Terminal. It is a globular crown of many short, crowded flower spikes.
Flower:	5 mm across. The sepals and petals are mild yellow. The dorsal sepal is erect with a short point. The lateral sepals are similar but slightly keeled, spreading. Mentum short. The petals are narrow. The lip is white with a purple-red partition. It has a rounded midlobe and cup-shaped base.

10. *Agrostophyllum stipulatum* (Griff.) Schltr. subsp. *stipulatum*

© The Tree Projects

Synonyms:
Appendicula stipulata Griff.
Appendiculopsis stipulata (Griff.) Szlach.

General Distribution:
Myanmar, Thailand, Peninsular Malaysia, Sumatra, Java, Borneo and Sulawesi.

Description:

Habitat and ecology:	An epiphyte or a lithophyte usually found growing on tree trunks, branches, twigs and is widely distributed from lowland to hilly areas.
Habit:	The stems are short and slender, 20–40 cm long.
Leaves:	Obtuse, arranged at right angles to the stem.
Inflorescence:	Terminal, usually 1-flowered.
Flower:	Wide opening, 3–5 mm across. The sepals and petals are white. The dorsal sepal is concave. Lateral sepals are broadly and obliquely triangular, forming a broad mentum. Petals are narrow. The white lip is saccate at its base, the midlobe is yellow, oblong-ovate at a right angle, and the side lobes are triangular. The column apex has incurved purple stelids on each side.
Flowering:	Throughout the year.

11. *Ania penangiana* (Hook.f.) Summerh.

Synonyms:
Tainia penangiana Hook.f.
Tainia taiwaniana S.S.Ying

General Distribution:
India, Vietnam, Thailand and
Peninsular Malaysia.

Description:

Habitat and ecology:	A terrestrial or a lithophyte which grows in leaf litter in hilly areas.
Habit:	It has ovoid pseudobulbs which are usually partially buried.
Leaves:	Pleated and palm-like, growing 40 cm long. Each pseudobulb has a single leaf, which is borne on a 20 cm long stalk.
Inflorescence:	Emerging from the underside of the bulb, growing 30–45 cm long. It is an erect stalk that bears 8–10 well-spaced flowers.
Flower:	Wide opening, about 2 cm across. Sepals and petals are golden brown with faint purple veins. Sepals are lanceolate while the petals are narrowly oblanceolate. The curved lip is white with a yellow midrib.
Flowering:	November–February.

Ania penangiana earns the specific epithet '*penangiana*' because the type specimen (as *Tainia penangiana* Hook.f.) was collected from Penang.

12. *Aphyllorchis pallida* Blume

Synonyms:
Aphyllorchis hasseltii Blume
Aphyllorchis gracilis Schltr.

General Distribution:
Thailand, Peninsular Malaysia,
Sumatra, Java, Borneo and
the Philippines.

Description:

Habitat and ecology:	A terrestrial holomycotroph which grows in accumulated leaf litter on the forest floor and is widely distributed from lowland to hilly areas.
Habit:	It has an underground rhizome with thick roots that send out slender stems which can grow to 50 cm long.
Leaves:	The plant has no leaves and does not photosynthesise, instead relying upon fungi for nutrition.
Inflorescence:	Terminal. The pale stalk is erect, bearing 10–25 dull flowers. It is usually indistinguishable from the surrounding leaf litter.
Flower:	Not opening widely. The sepals and petals are pale yellow with a distinct longitudinal purple midrib. Dorsal sepal hooded over column, lanceolate, apex obtuse to subacute. Lateral sepals falcate, apex acute. Petals obliquely ovate-elliptic, apex obtuse-rounded. The trilobed lip at an acute angle from column axis, saccate at base, apex thick and fleshy, lateral margins curved upwards forming a concave lamina. It is dull yellow with purple streaks. Column is white and broadly winged at the apex.

13. *Appendicula pendula* Blume

Synonyms:
Appendicula lancifolia Hook.f.
Appendicula maingayi Hook.f.

General Distribution:
Thailand, Peninsular Malaysia,
Sumatra, Java, Borneo, the Philippines
and Sulawesi.

Description:

Habitat and ecology:	An epiphyte, lithophyte or a terrestrial which grows on the lower section of trees and near mossy stream banks.
Habit:	The stem is pendulous, growing to 100 cm long.
Leaves:	Slightly oval-shaped, tapering at an acute tip and up to 12 cm long (usually much smaller).
Inflorescence:	Terminal, growing to 15 cm long. It is pendulous, developing from the end of the stem carrying many flowers which are produced in succession over a long period of time.
Flower:	Not opening widely, 7–10 mm across. Sepals and petals are greenish yellow. Dorsal sepal concave and broadly pointed, while the lateral sepals are obliquely triangular. Petals ovate, obtuse. The fleshy lip is white with a red blotch in the inner part. It is bent in the middle.

14. *Arundina graminifolia (D.Don) Hochr. subsp. graminifolia*

Common Name:
Bamboo Orchid

Synonyms:
Arundina chinensis Blume
Arundina speciosa Blume

General Distribution:
India, Bangladesh, Thailand, Peninsular Malaysia, Java, Sumatra and Borneo.

Description:

Habitat and ecology:	A terrestrial growing in sunny areas like road cuttings and in secondary growth (*belukar*). It is widely distributed from lowland to hilly areas.
Habit:	The stem is erect and bamboo-like, up to 2.5 m in height.
Leaves:	Narrow, grass-like and keeled, 10–20 cm long.
Inflorescence:	Terminal, sometimes branched. It is erect and elongates gradually by producing a succession of flowers, normally one or two at a time.
Flower:	Wide opening, about 5 cm across. Sepals and petals are pinkish-white. Sepals are narrowly elliptic while petals are broadly elliptic. The trumpet-shaped lip is bright purple with a yellow blotch in the centre. However, the flower displays much variation in colour and size across its wide range.[24]
Flowering:	Throughout the year.

Its rhizomes are believed to have antibacterial properties which can be used for controlling bacterial infections.[25] A decoction of the roots is used to control hepatitis and jaundice.[26] In East Malaysia, the flowers of the highland variety are stir-fried and eaten as it is believed the bitter-tasting flowers can help relieve high blood pressure.[27]

15. *Bromheadia finlaysoniana* (Lindl.) Miq.

Common Name:
Pale Reed Orchid

Synonyms:
Grammatophyllum finlaysonianum Lindl.
Bromheadia palustris Lindl.

General Distribution:
Thailand, Peninsular Malaysia,
Sumatra, Borneo, New Guinea
and Australia.

Description:

Habitat and ecology:	A terrestrial. It is quite easily spotted in secondary growth (*belukar*) and is widely distributed from lowlands to hilly areas.
Habit:	It has a creeping underground rhizome with closely-set erect stems up to 1–2 m long.
Leaves:	Elliptic, about 12 cm long.
Inflorescence:	Terminal, rarely branched, growing to more than 10 cm. It is erect and can produce about 50 flowers which develop in succession. At a given time one or two flowers bloom on each stalk.
Flower:	Wide opening, about 5 cm across, fragrant. The sepals and petals are creamy white. The sepals are ovate-elliptic while the petals are ovate to obovate. The trilobed lip has a warty yellow midlobe while the side lobes have purple veins. The column is white. Flowers last less than one day.
Flowering:	Throughout the year. Gregarious flowering is triggered by rain.

It is commonly seen in hilly areas of Penang. It used to be a popular garden plant which was hardy and suitable for borders or flowerbeds.[28] The roots are boiled and drunk to treat rheumatism and joint pain.[29] In Sarawak, the Kelabit consume the flowers raw and as a cooked vegetable.[30] The flower stalks are believed to be an effective treatment for asthma.[31]

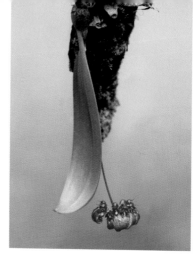

16. *Bulbophyllum corolliferum* J.J.Sm.

Synonyms:

Bulbophyllum curtisii (Hook.f.) J.J.Sm.
Cirrhopetalum curtisii Hook.f.

General Distribution:

Thailand, Peninsular Malaysia,
Sumatra and Borneo.

Description:

Habitat and ecology:	An epiphyte which grows in areas of light shade and high humidity.
Habit:	It has garlic-shaped ovoid pseudobulbs up to 1.5 cm long. The pseudobulbs are well-spaced on the rhizome.
Leaves:	Blunt, slightly bilobed apex, growing to 16 cm long. Each pseudobulb supports a single leaf.
Inflorescence:	Arising from the base of the pseudobulb, growing to 10 cm long. Flowers are arranged as an umbel at the tip, carrying 8–12 flowers.
Flower:	Entirely dark purple. The dorsal sepal is hooded with a slender tip. It is fringed with short purple hairs at the margins. The lateral sepals are distinct, joined from near the base to the tip, forming a curved pouch. The petals are a little shorter than the dorsal sepal and fringed with short purple hairs at the margins. The lip is purple-yellow and curved in the middle. The column is cream-pale yellow at the basal half and yellow at the apical half; the apex has short stelids and is broadly winged.

17. *Bulbophyllum depressum* King & Pantl.

Synonyms:
Bulbophyllum acutum J.J.Sm.
Bulbophyllum hastatum Tang & F.T.Wang

General Distribution:
Bhutan, India, Thailand,
Peninsular Malaysia, Sumatra, Java and Borneo.

Description:

Habitat and ecology:	An epiphyte or a lithophyte which grows on granite boulders in hilly areas.
Habit:	It has flattened and ovoid pseudobulbs which can grow 8 mm long. The pseudobulbs are well spaced on the rhizome.
Leaves:	Small and thick, about 6 mm long. Each pseudobulb supports a single leaf.
Inflorescence:	Arising on the rhizome nodes between the pseudobulbs, growing 5 mm long. It is 1-flowered.
Flower:	Multi-coloured. The sepals and petals have a yellow tip, a maroon mid-section and a white basal section. The dorsal sepal is narrowly ovate. The lateral sepals are narrowly ovate and slightly longer than the dorsal sepal. The petals are elliptic with an acute apex. The trilobed lip is maroon.
Flowering:	May.

18. *Bulbophyllum gracillimum* (Rolfe) Rolfe

Common Name:
Wispy Umbrella Orchid

Synonyms:
Bulbophyllum leratii (Schltr.) J.J.Sm.
Bulbophyllum psittacoides (Ridl.) J.J.Sm.

General Distribution:
Thailand, Peninsular Malaysia, Sumatra, Borneo, Maluku, Sulawesi and Australia.

Description:

Habitat and ecology:	An epiphyte or a lithophyte which grows in hilly areas on large granite rocks and tree trunks.
Habit:	It has ovoid pseudobulbs which grow to 2 cm long. In younger plants, pseudobulbs are closely arranged on the rhizome and become more spaced out as the plant matures.
Leaves:	Elliptic-oblong, growing to 12 cm long. Each pseudobulb supports a single leaf.
Inflorescence:	Arising from the base of the pseudobulb, growing to 30 cm long. Flowers form an umbel at the end of the spike. Each spike carries about 10 flowers.
Flower:	Sepals and petals are yellow, lightly flushed red at the base and become more intensely red towards the tips. The dorsal sepal is hooded with a slender tail at the apex and short hairs at the margins. The lateral sepals are joined at the base and become long curved tails. Petals are narrow and similar in length to the dorsal sepal, with a slender tail and short hairs at the margins. The lip is white, flushed purple-red at the basal half and curved in the middle. Column is yellow, flushed purple-red at the base.
Flowering:	November–December.

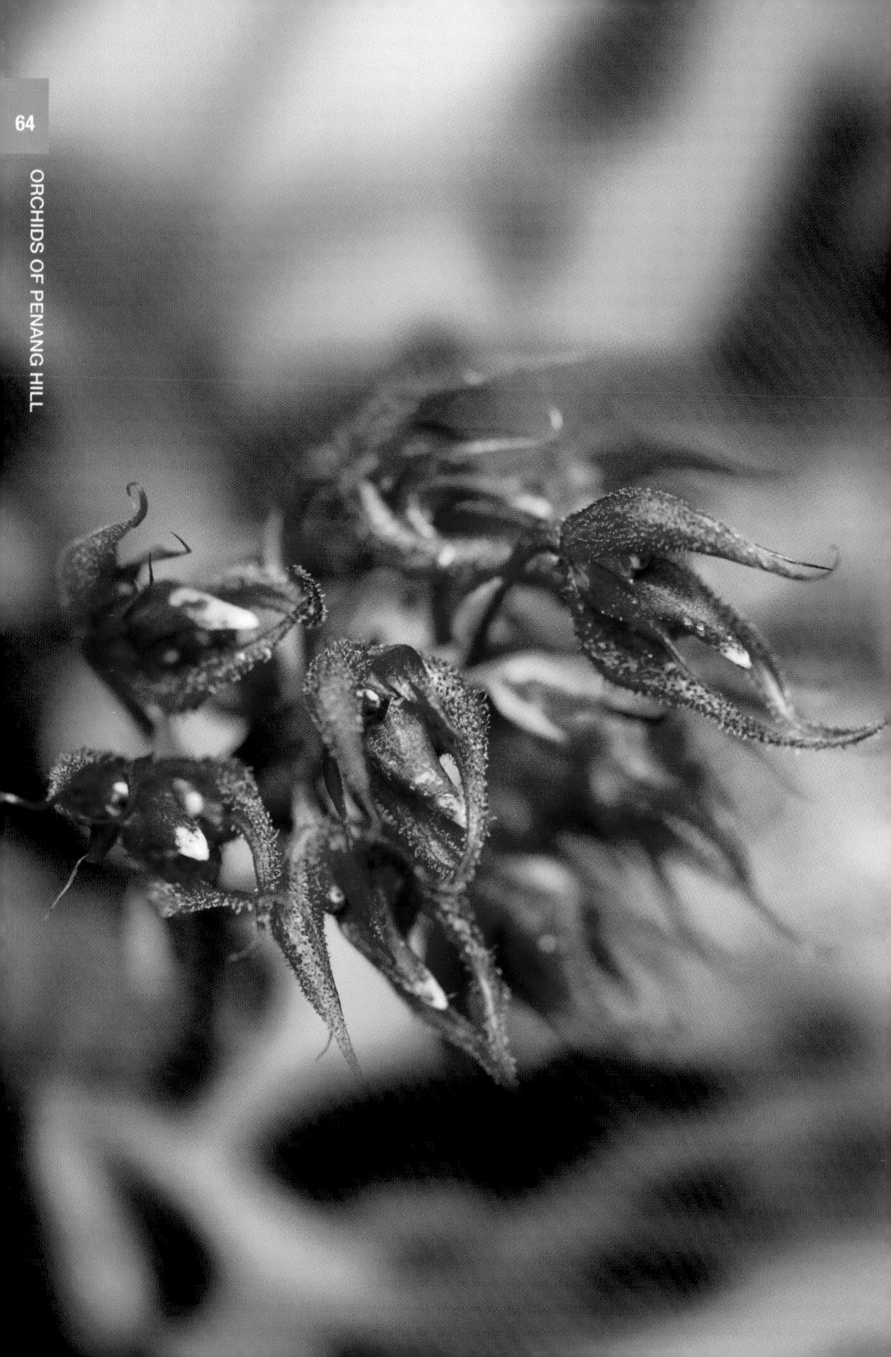

19. *Bulbophyllum lasianthum* Lindl.

Synonyms:
Anisopetalon lasianthum Kuhl ex Hook.f.
Phyllorkis lasiantha (Lindl.) Kuntze

General Distribution:
Peninsular Malaysia, Sumatra, Java,
Borneo, Maluku and Sulawesi.

Description:

Habitat and ecology:	An epiphyte or a lithophyte distributed from lowland to hilly areas.
Habit:	The plant has slightly flattened pseudobulbs which are 7.5 cm long. The pseudobulbs are well spaced on the rhizome.
Leaves:	Large, growing to 40 cm long. Each pseudobulb supports a single leaf.
Inflorescence:	Arising from the base of the pseudobulb, growing to about 10 cm long. It is stout, covered in numerous sheaths and carries about 12 closely-arranged flowers.
Flower:	Not opening widely, foul-smelling. The sepals and petals are purple-red. The dorsal sepal is ovate and concave while the lateral sepals are slightly narrower. The sepals are hairy. The petals are narrow, pointed and glabrous. The fleshy lip is purple red with a blotch of white at the tip. It is strongly curved near the base. Column is purple red with a white blotch at the front.

The carrion smell and dark colour of its flowers are known to attract blowflies which pollinate the plant. In 2010, researchers observed how a swarm of flies crawling over the inflorescence removed all the pollen in an hour.[32]

20. *Bulbophyllum medusae* (Lindl.) Rchb.f.

Common Name:
Medusa Orchid

Synonyms:
Cirrhopetalum medusae Lindl.
Phyllorkis medusae (Lindl.) Kuntze

General Distribution:
Thailand, Peninsular Malaysia, Sumatra, Borneo and Lesser Sunda Islands.

Description:

Habitat and ecology:	An epiphyte commonly found in lowland forests and sometimes in urban areas.
Habit:	It has pseudobulbs which are 2–3.5 cm long and are well spaced on the rhizome.
Leaves:	Stout, growing 10–20 cm long. Each pseudobulb supports a single leaf.
Inflorescence:	Arising from the base of the pseudobulb, growing 10 cm long. It is erect to horizontal, supporting a dense umbel of about 15 flowers.
Flower:	The sepals and petals are white with a tinge of yellow near the base, sometimes spotted purple. The dorsal sepal is ovate with a long slender tail. The lateral sepals are similar to the dorsal sepal but with much longer tails. The petals are ovate-triangular and small. The small lip is shiny yellow with blotches of red at the base. It is recurved, pointed at the apex. Column is yellow with red blotches ventrally.

The thread-like sepals which drop out freely from the umbel inflorescence give it the appearance of the Greek mythological monster Medusa; hence its name.[33]

21. *Bulbophyllum pileatum* Lindl.

Synonyms:
Phyllorkis pileata (Lindl.)
Sarcopodium pileatum (Lindl.) Lindl.

General Distribution:
Peninsular Malaysia, Sumatra and Borneo.

Description:

Habitat and ecology:	An epiphyte which is widely distributed from mangrove swamps to hilly areas.
Habit:	Pseudobulbs are ovoid, growing to 2.5 cm long. They are well spaced on the rhizome. There are stiff bristles around the base of the pseudobulb.
Leaves:	Oblong-elliptical, growing to 10 cm long. Each pseudobulb supports a single leaf.
Inflorescence:	Arising from the base of the pseudobulb, growing to 6 cm long. It is 1-flowered.
Flower:	Wide opening, large. The sepals and petals are dull yellow. The dorsal sepal is ovate, erect to pointing forward. The lateral sepals are obliquely oblong-ovate and curve inwards. Petals are ovate-elliptic, smaller than the sepals. The lip has a darker tone of yellow with purple lines at the base and is fleshy, narrowly ovate, blunt at the tip. The slender column is yellow.
Flowering:	April–May.

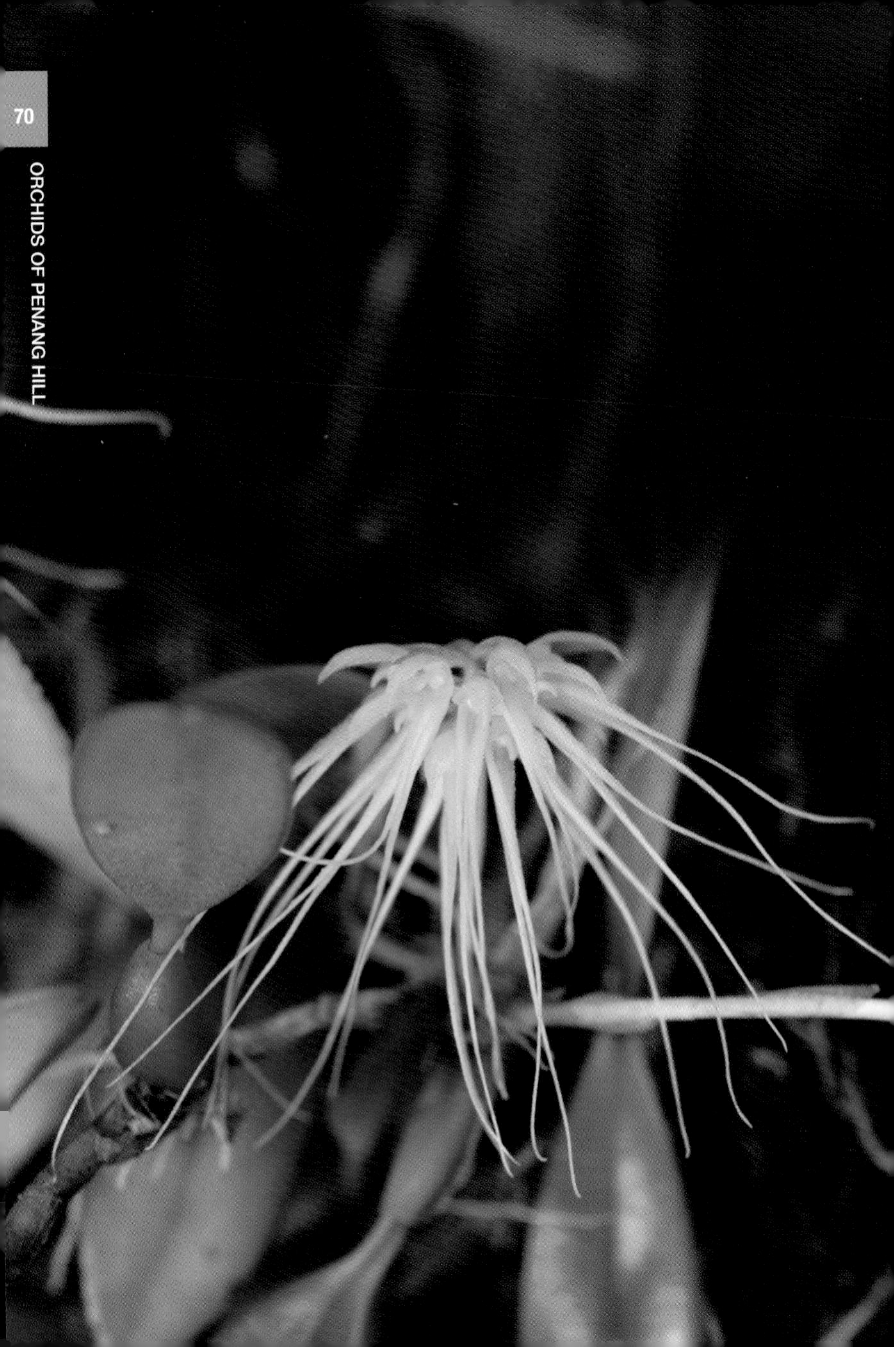

22. *Bulbophyllum vaginatum* (Lindl.) Rchb.f.

Common Name:
Magrah Batu

Synonyms:
Bulbophyllum whiteanum (Rolfe) J.J.Sm.
Cirrhopetalum whiteanum Rolfe

General Distribution:
Thailand, Peninsular Malaysia, Sumatra, Bangka, Java, Borneo and Maluku.

Description:

Habitat and ecology:	An epiphyte which grows in freshwater swamps, lowland forest and even in urban areas, on the branches of large trees. It is widely distributed but primarily in lowland and lower parts of hilly areas.
Habit:	The ovoid pseudobulbs are 2 cm long, well-spaced on the tough rhizome.
Leaves:	Stiff, thick and oblong growing to 12 cm long. Each pseudobulb supports a single leaf.
Inflorescence:	Arising from the base of the pseudobulb, growing to 10 cm long. Flowers form an umbel at the end of the spike, bearing up to 15 flowers.
Flower:	The sepals and petals are pale yellow. The dorsal sepal has short-fringed edges and curves forward. The lateral sepals are also fringed and have long tails. The petals are small and fringed along the margins. The fleshy lip has a brighter yellow tone and is recurved towards the tip.
Flowering:	April–May.

The roasted fruit extract can be used to treat earaches.[34]

23. *Calanthe pulchra* (Blume) Lindl.

Synonyms:
Styloglossum pulchrum (Blume)
   T.Yukawa & P.J.Cribb
Amblyglottis pulchra Blume

General Distribution:
India, Thailand, Peninsular Malaysia,
Sumatra, Java, Borneo and the Philippines.

Description:

Habitat and ecology:	A terrestrial or a lithophyte which prefers damp areas and grows abundantly along slopes near stream valleys and occasionally inside trickling streams. It is widely distributed from lowland to hilly areas.
Habit:	The plant grows to 75 cm long. It has a small pseudobulb.
Leaves:	Elliptic, growing to about 70 cm long.
Inflorescence:	Arising from the base of the plant. It is erect, growing to 60 cm tall and densely crowded with 40–60 flowers.[35]
Flower:	Not opening widely, about 1.5 cm across. The sepals and petals are orange. The dorsal sepal is ovate-elliptic and narrowly acute at the apex. The lateral sepals are slightly smaller. The petals are elliptic-lanceolate and wider than the lateral sepals. The trilobed lip has a hooked spur. The midlobe is orange and the side lobes, red.
Flowering:	August–October.

24. *Coelogyne rochussenii* de Vriese

Common Name:
Necklace Orchid

Synonyms:
Coelogyne macrobulbon Hook.f.
Pleione macrobulbon (Hook.f.) Kuntze

General Distribution:
Thailand, Peninsular Malaysia, Sumatra, Java, Borneo, the Philippines, Maluku and Sulawesi.

Description:

Habitat and ecology:	An epiphyte or a lithophyte which usually grows on trees overhanging streams and valleys.
Habit:	It has stout, cylindrical pseudobulbs which taper towards the top. They grow to 20 cm long.
Leaves:	Plicate, growing to 30 cm long. Each pseudobulb supports two leaves.
Inflorescence:	Emerging near the base of the pseudobulb, growing to 50 cm long. It is pendulous, carrying up to 40 flowers.
Flower:	Not opening widely, fragrant. The sepals and petals are pale greenish-yellow. The sepals are narrowly elliptic with an acute apex. The petals are oblanceolate to spathulate with an acute apex. The trilobed lip is white, the side lobes flushed pale brown with white veins, the midlobe white, the base sometimes flushed pale yellow, keels white, apex marked yellow. The column is white.
Flowering:	October–December.

This species is a favourite among hobbyists due to its attractive flowers and relative ease of growing. Avid hobbyists grow the plant in hanging circular pots; with the right care, the plant is able to produce dozens of spikes with hundreds of flowers blooming at once. The flower spikes which dangle down make it look akin to a flower basket!

25. *Cymbidium dayanum* Rchb.f.

Common Name:
Phoenix Orchid

Synonyms:
Cymbidium acutum Ridl.
Cymbidium leachianum Rchb.f.

General Distribution:
China, Japan, India, Myanmar, Laos, Thailand, Peninsular Malaysia, Borneo and Maluku.

Description:

Habitat and ecology:	An epiphyte or a terrestrial usually found in hilly areas, growing in medium-sized clumps.
Habit:	It has short pseudobulbs, which are closely arranged, forming clumps.
Leaves:	Long, narrow and grass-like, growing to 50 cm long; each pseudobulb bears about 5 leaves.
Inflorescence:	Emerging from the base of the pseudobulb, growing to 20 cm long. It is horizontal to pendulous, supporting 10–17 flowers.
Flower:	The sepals and petals are off-white with a longitudinal purple band in the middle. The sepals and petals are narrowly lanceolate-elliptic; the petals are slightly shorter. The petals usually face forward and inwards at the tip over the column. The lip is maroon with a yellow blotch at the midlobe. The lip is trilobed, with the midlobe curved downwards and backwards. The column is maroon, the anther dull yellow.
Flowering:	August–September.

This species is popular among orchid lovers as it blooms profusely and is easy to care for. Other varieties of this species produce red flowers while certain varieties have distinctly variegated leaves.

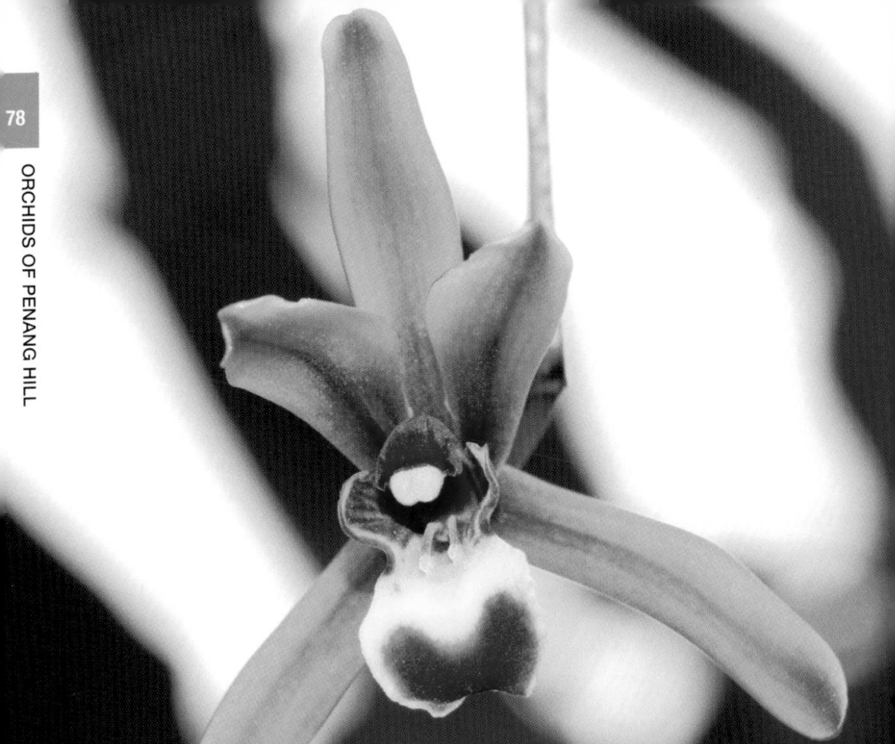

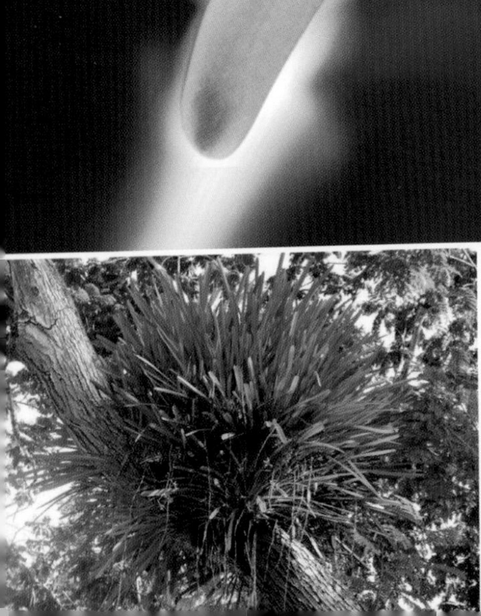

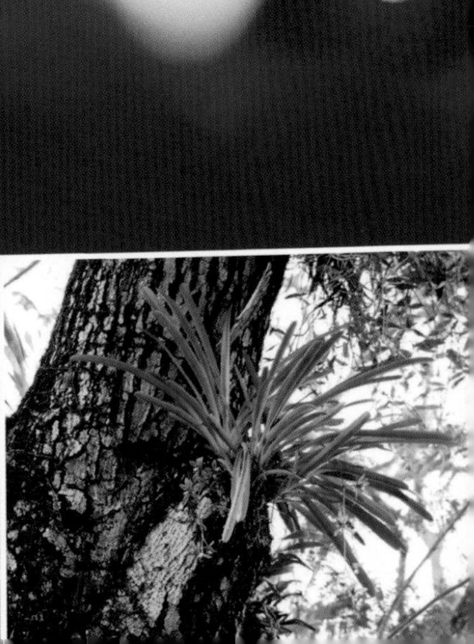

26. *Cymbidium finlaysonianum* Lindl

Common Name:
Bunga Chandarek

Synonyms:
Cymbidium tricolor Miq.
Cymbidium wallichii Lindl.

General Distribution:
Thailand, Peninsular Malaysia, Sumatra, Java, Borneo, the Philippines, Sulawesi and Maluku.

Description:

Habitat and ecology:	An epiphyte or a lithophyte widely distributed from the seaside to hilly areas, growing between crotches of trees or on granitic rocks as heavy clumps. In urban areas of Penang it can be seen forming huge clumps on large roadside trees.
Habit:	It has short pseudobulbs, which are closely arranged, forming dense clumps.
Leaves:	Thick and stiff, growing to 75 cm long. Each pseudobulb supports about 5 leaves.
Inflorescence:	Emerging from the base of the pseudobulb, growing to 100 cm long. It is horizontal to pendulous, with up to 25 well-spaced flowers.
Flower:	Wide opening, mildly fragrant. The sepals and petals are greenish-yellow with a longitudinal purple band running through the middle. The sepals and petals are oblong, with the petals facing forward at an acute angle over the column. The lip has a white midlobe which is curved downwards and inwards at the apex. The lip's midlobe has a purple crescent-shaped marking in the middle near the apical margins, the side lobes are white with purple veins, the lip disc is white, the basal half has a median longitudinal purple band, above which is a yellow longitudinal band extending to about halfway on the midlobe, keels purple. Column is yellow, densely marked purple, the anther bright yellow.
Flowering:	Throughout the year.

It is possibly one of the most common wild orchids in Penang. In the past, when spirits were believed to cause illnesses, traditional medicine men in Peninsular Malaysia used it to remove 'bewitchment' from patients.[36]

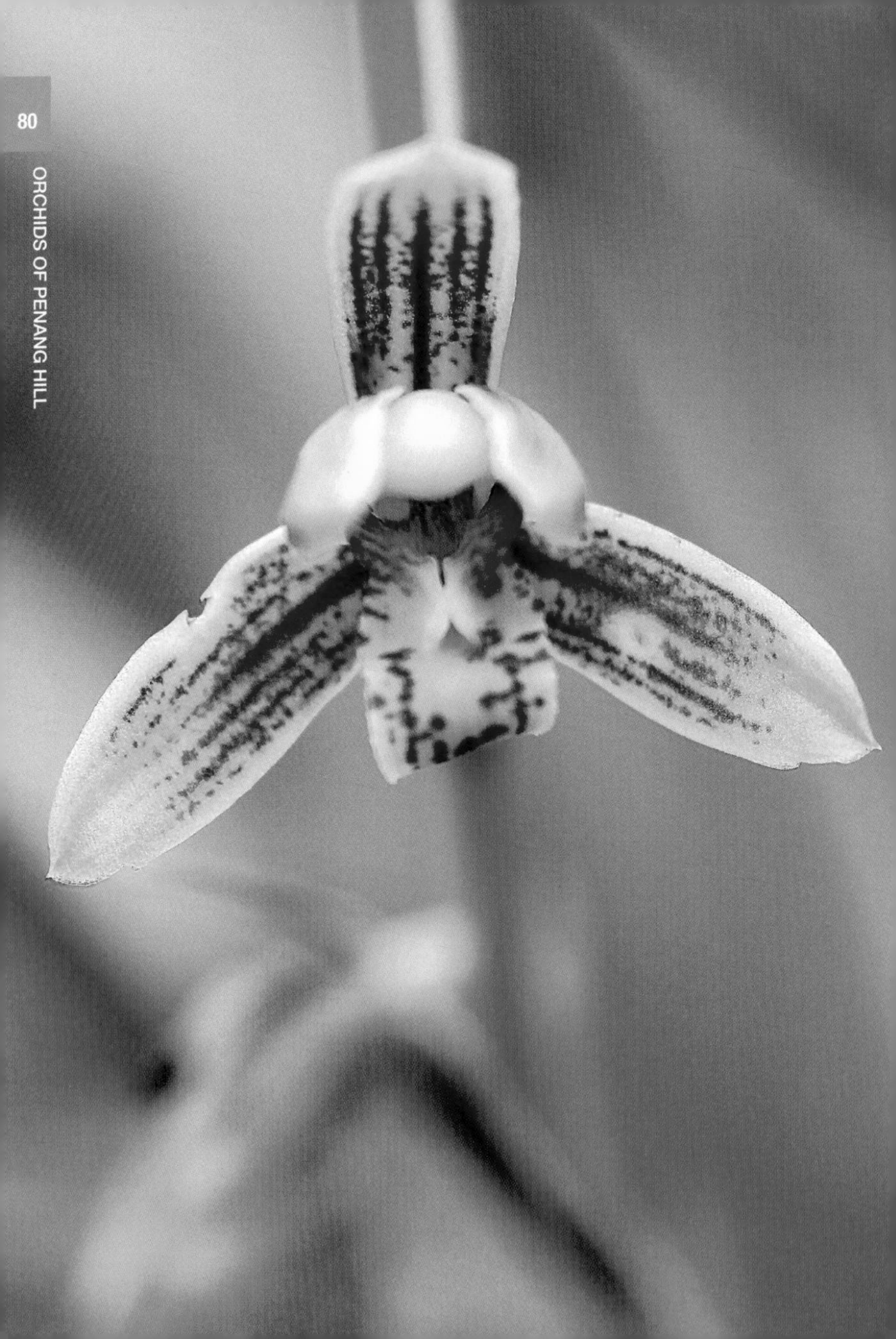

27. *Cymbidium haematodes* Lindl.

Synonyms:
Cymbidium ensifolium var. *haematodes*
  (Lindl.) Trimen
Cymbidium siamense Rolfe ex Downie

General Distribution:
China, India, Laos, Thailand,
Peninsular Malaysia, Sumatra, Java,
Borneo and Sulawesi.

Description:

Habitat and ecology:	A terrestrial which grows in the undergrowth on slopes, primarily in hilly areas.
Habit:	It has small, ovoid pseudobulbs which are usually partially buried.
Leaves:	Strap-shaped, arching near the tip, growing to 90 cm long.
Inflorescence:	Arising from the base of the pseudobulb, growing to 60 cm long. The spike bears 4–8 flowers.
Flower:	Not opening widely, horizontal or slightly nodding. Sepals and petals are yellow, irregularly streaked red. Sepals and petals are slightly oblong-lanceolate, apex acute. The lip is yellow with irregular red blotches or markings. The lip is trilobed, with the midlobe curved down at the front. The disc has two ridges at its base which extend to slightly above the base of the midlobe and converge about midway to form a short tube.
Flowering:	April–May.

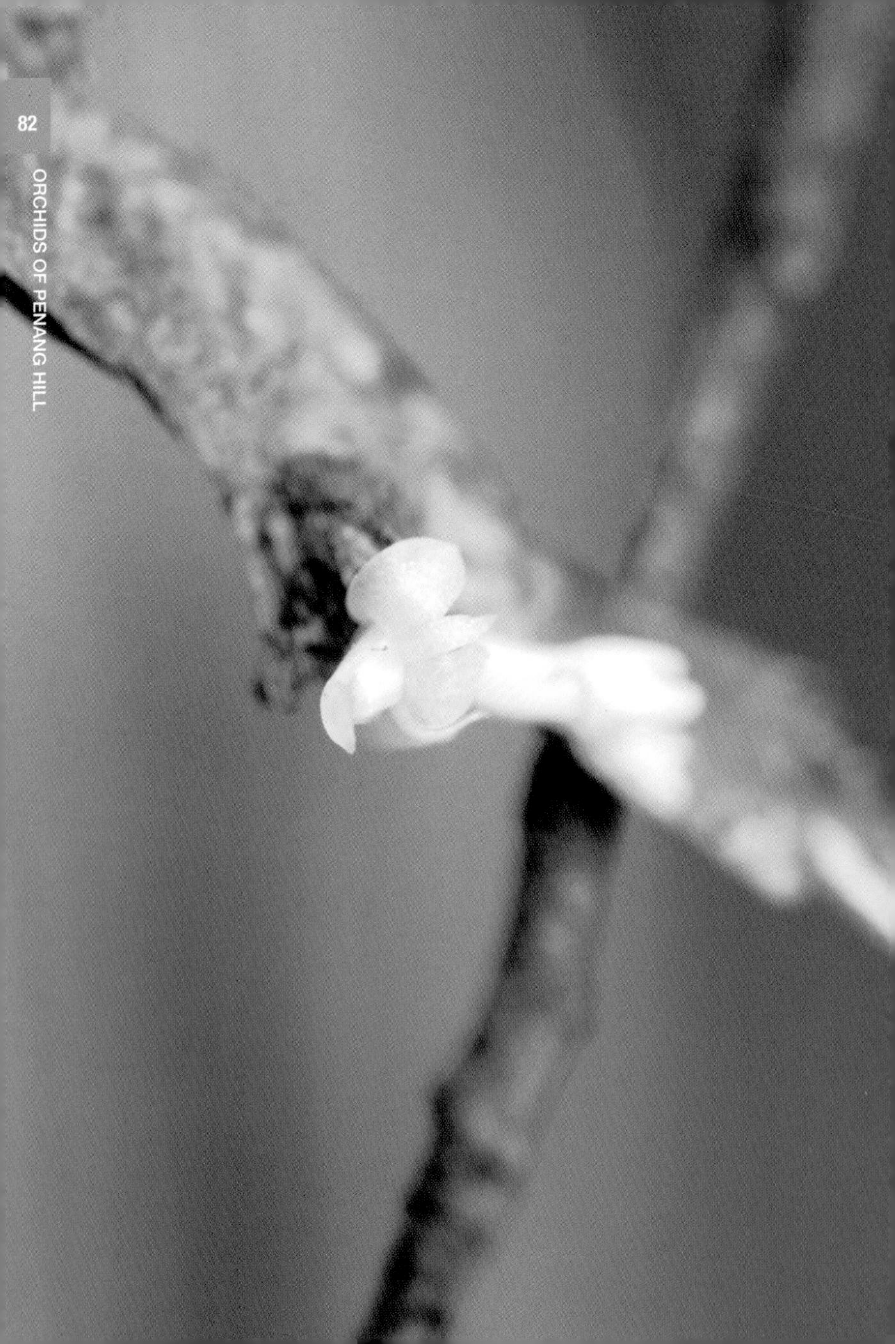

28. *Dendrobium aloifolium* (Blume) Rchb.f.

Synonyms:
Dendrobium lobbii Lindl.
Dendrobium serra (Lindl.) Lindl.

General Distribution:
Myanmar, Cambodia, Thailand,
Peninsular Malaysia, Sumatra, Java,
Borneo and the Philippines.

Description:

Habitat and ecology:	A widespread epiphyte usually found growing on tree trunks in hilly areas.
Habit:	The stem grows to 45 cm long. It initially grows upright, and slowly bends down to a pendulous form.
Leaves:	Obliquely triangular, apex acute, flattened, fleshy.
Inflorescence:	Produced along the apical portion as a pseudoraceme. It grows to 15 cm long, and flowers successively.
Flower:	Wide opening. The sepals and petals are white, reflexed. Dorsal sepal ovate-elliptic, apex acute. Lateral sepals obliquely oblong-lanceolate, free part obliquely triangular, apex acute. Petals narrowly elliptic, apex subacute. The white lip curves upwards at the mid-section. The anther can be off-white, pale yellow or green.

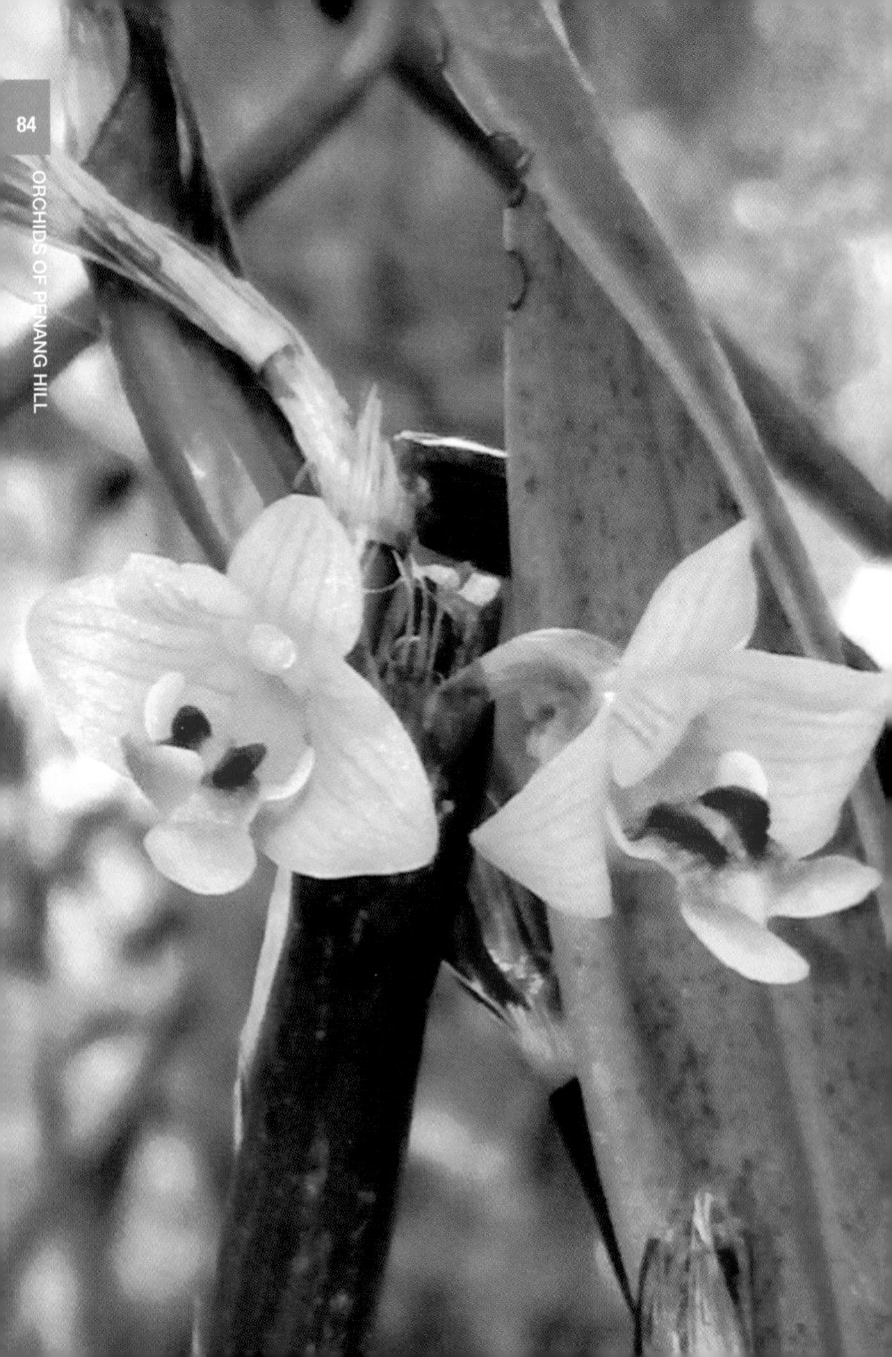

29. Dendrobium *angustifolium* (Blume) Lindl. subsp. *angustifolium*

Synonyms:
Dendrobium bancanum J.J.Sm.
Flickingeria bancana (J.J.Sm.) A.D.Hawkes

General Distribution:
India, Vietnam, Thailand,
Peninsular Malaysia, Sumatra and Borneo.

Description:

Habitat and ecology:	An epiphyte widely distributed from lowland to hilly areas, usually growing on tree trunks.
Habit:	It has spindle-shaped yellowish green pseudobulbs that grow to 3 cm long.
Leaves:	Narrowly lanceolate. Each pseudobulb supports a single leaf.
Inflorescence:	Produced near the pseudobulb apex on the abaxial. It is usually 1-flowered (occasionally two).
Flower:	Wide opening. The sepals and petals spreading, often curved backwards. The sepals and petals are dull yellow with light maroon veins. The lateral sepals are white, sometimes tinged purple-pink on the inner margins. The lip is trilobed, the midlobe is bilobulate, yellow, the base flushed purple-pink, the disc with 3 keels; the median keel is white, the lateral keels are dark purple. The column is pale greenish-yellow, anther pale yellow.

30. *Dendrobium anosmum* Lindl.

Common Name:
Purple Rain Orchid

Synonyms:
Callista anosma (Lindl.) Kuntze
Dendrobium macranthum Hook.

General Distribution:
Myanmar, Thailand, Peninsular Malaysia,
Sumatra, Java, Borneo, Maluku and
New Guinea.

Description:

Habitat and ecology:	An epiphyte widely distributed from lowland to hilly areas growing on tree trunks and branches.
Habit:	It has stout pendant stems which grow to 120 cm long or more.
Leaves:	Elliptic. The plant is deciduous and older stems often shed leaves during the dry season.
Inflorescence:	Arising from the internodes on the stem. The number of flowers on each stem can vary from a single flower to about a dozen.
Flower:	6–10 cm across, fragrant with a raspberry-like smell. They have mauve-purple elliptic sepals and petals. The petals are broader than the sepals. The hairy lip is funnel-like, mauve-purple with a dark purple blotch in the middle. The column is dark purple.
Flowering:	February–April.

Ironically, the epithet '*anosmum*' means 'unscented'. It requires a distinct dry season to flower well.[37]

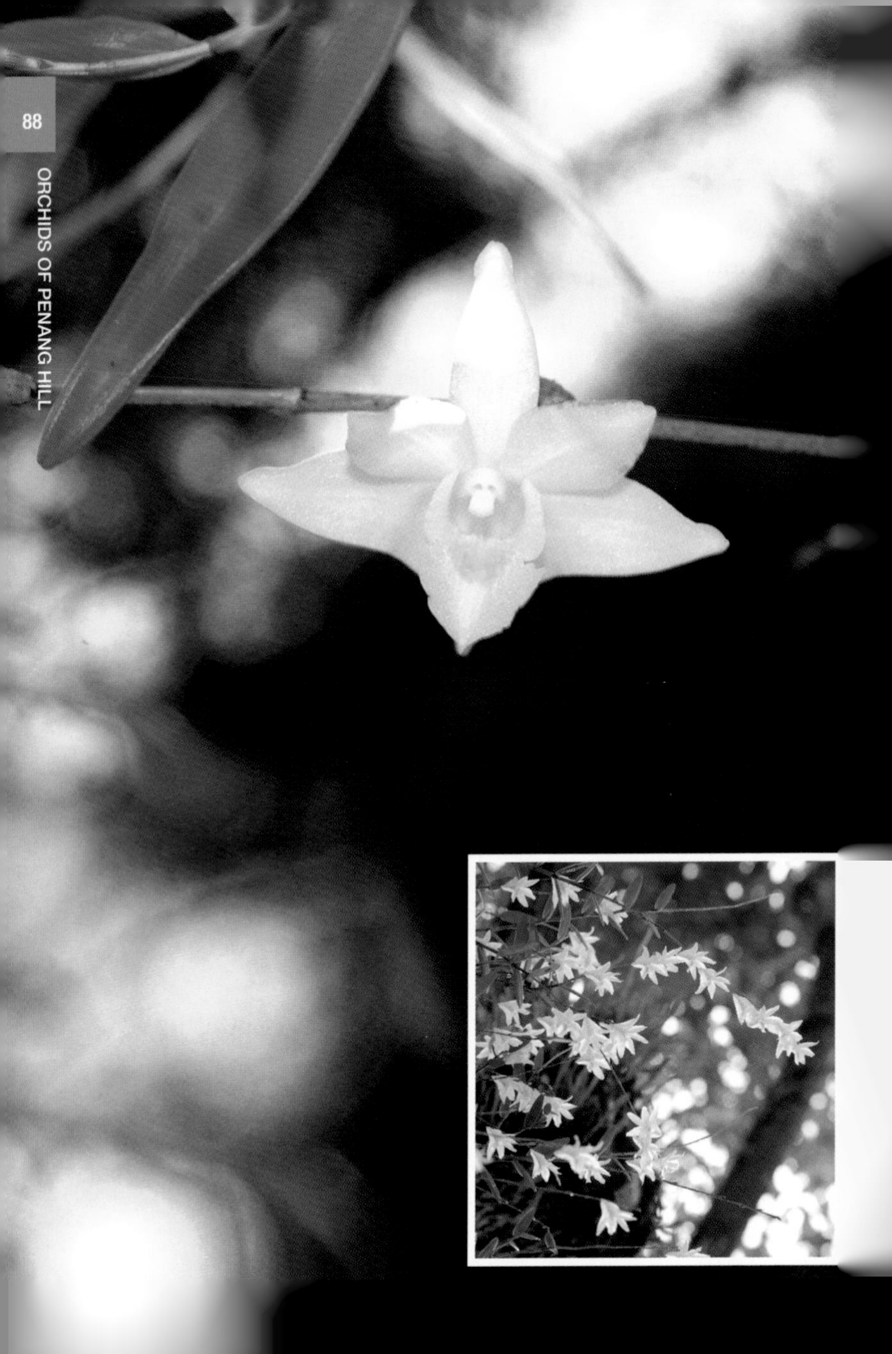

31. *Dendrobium crumenatum* Sw.

Common Name:
Pigeon Orchid

Synonyms:
Epidendrum caninum Burm.f.
Onychium crumenatum (Sw.) Blume

General Distribution:
India, Sri Lanka, Thailand,
Peninsular Malaysia, Sumatra,
Borneo and the Philippines.

Description:

Habitat and ecology:	An epiphyte or a lithophyte widely distributed from lowland to hilly areas. It usually grows on tree trunks and branches, granitic boulders and even rooftops in urban areas.
Habit:	The stems grow to 70 cm long and are basally swollen.
Leaves:	Thick and leathery.
Inflorescence:	Borne along the distal portion of the stem. The flowers are borne singly from tufts of dry, chaffy bracts.
Flower:	Wide opening, about 3 cm across, fragrant. The sepals and petals are white. The dorsal sepal is elliptical. The lateral sepals are obliquely triangular. The petals are elliptical. The lip is white with a prominent yellow patch in the middle, the disc with 5 warty keels from the base extending to near the apex. The column and anther are white. Flowers only last one day.
Flowering:	No distinct season. Gregarious flowering is triggered by rainstorms that result in a temperature drop of about 5.6°C. The flowers bloom about 9 days later.

It is a very common species in urban areas of Penang. The juice from the pseudobulbs is used to treat earache and rheumatism.[38] In the Philippines, the stem fibres are used as braiding materials.[39] The flowers are pollinated by the Asiatic honeybee, *Apis cerana indica*.

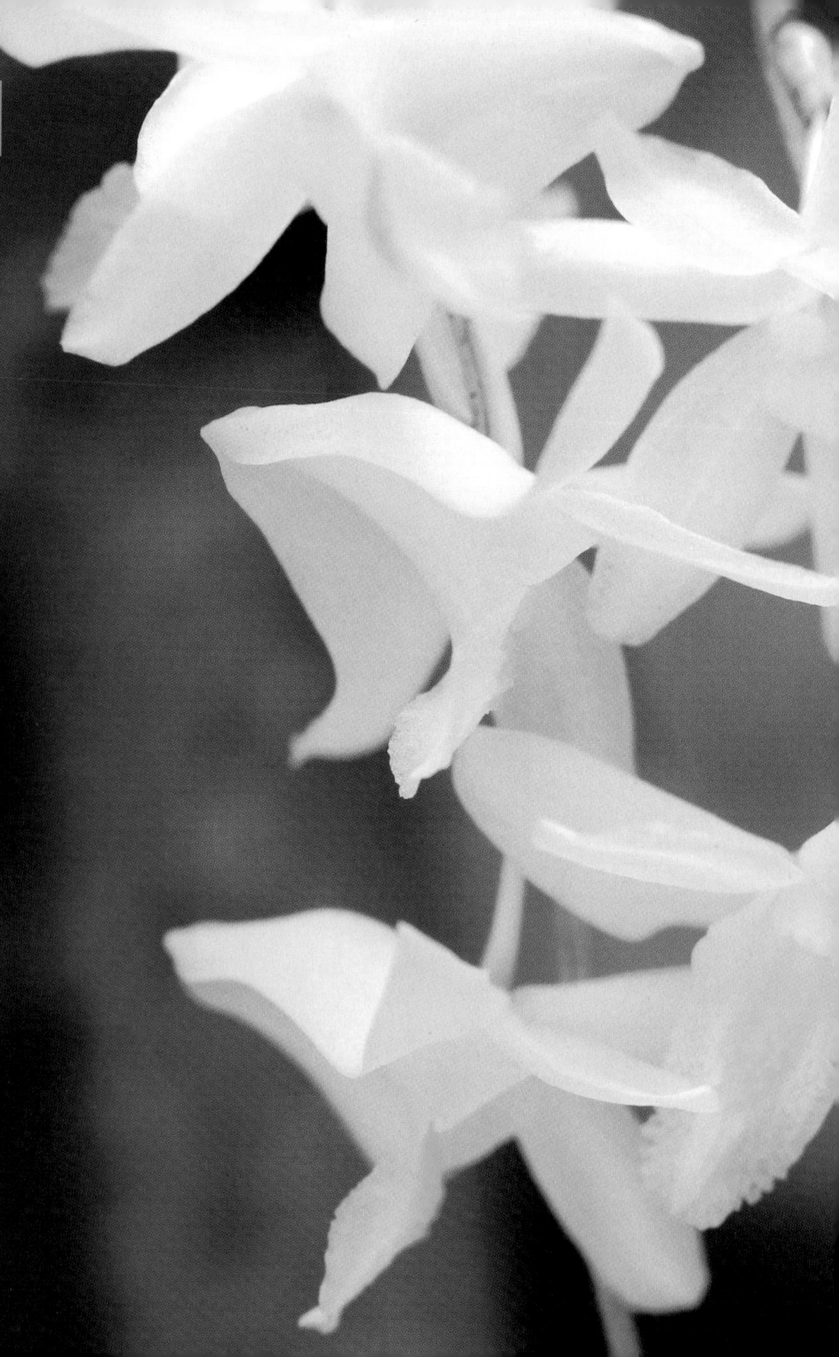

Dendrobium crumenatum Sw

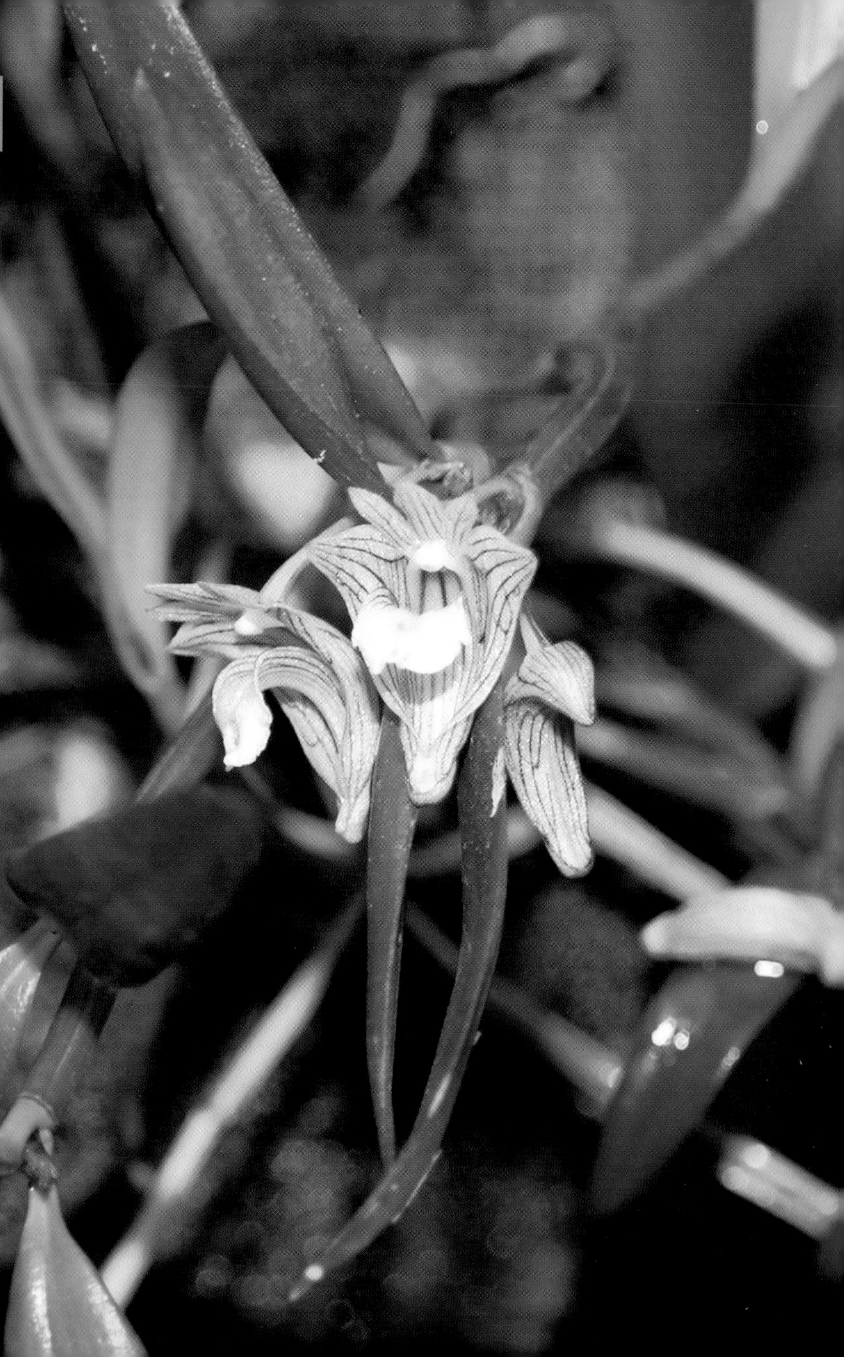

32. *Dendrobium pachyphyllum* (Kuntze)Bakh.f.

Synonyms:
Dendrobium pumilum Roxb.
Callista pachyphylla Kuntze

General Distribution:
India, Myanmar, Thailand, Peninsular Malaysia, Sumatra, Java and Borneo.

Description:

Habitat and ecology:	An epiphyte found in hilly areas, growing on trees in sunny spots, where it forms dense mats.
Habit:	The stems grow to 50 mm long, with a few internodes. The upper stem internode is swollen.
Leaves:	Fleshy, growing 9–55 mm long. Each pseudobulb carries 2 leaves.
Inflorescence:	Arising at the apical portion of the stem carrying one or two flowers.
Flower:	Wide opening, fragrant. The sepals and petals are cream-pale yellow with purple veins. The dorsal sepal is oblong-triangular. The lateral sepal is obliquely triangular. Petals are small and linear to narrowly obovate. The lip is cream-pale yellow with purple veins at the distal two-thirds near the margins, the lamina has a yellow blotch near the apex. The lip is curved at the proximal two-thirds. Flowers only last one day.

The root decoction is used to treat oedema.[40]

33. *Dienia ophrydis* (J.Koenig) Seidenf.

Common Name:
Common Snout Orchid

Synonyms:
Malaxis ophrydis (J.Koenig) Ormerod
Crepidium ophrydis (J.Koenig) M.A.Clem.
& D.L.Jones

General Distribution:
China, Vietnam, Thailand,
Peninsular Malaysia, Sumatra, Java, Borneo and Maluku.

Description:

Habitat and ecology:	A terrestrial mainly found in hilly areas.
Habit:	It has fleshy pseudobulbs which grow to 20 cm long.
Leaves:	Spade-shaped. Each pseudobulb carries 3–5 leaves.
Inflorescence:	Terminal, growing to 30 cm long. The erect stalk is crowded with hundreds of closely-arranged flowers.
Flower:	Non-resupinate (appear upside down). The sepals and petals are light red or purplish red. The dorsal sepal and petals are oblong, the lateral sepals are obliquely ovate-elliptic. The trilobed lip is maroon, the apex paler. Its midlobe is narrow, the side lobes broad and blunt, auricles absent.

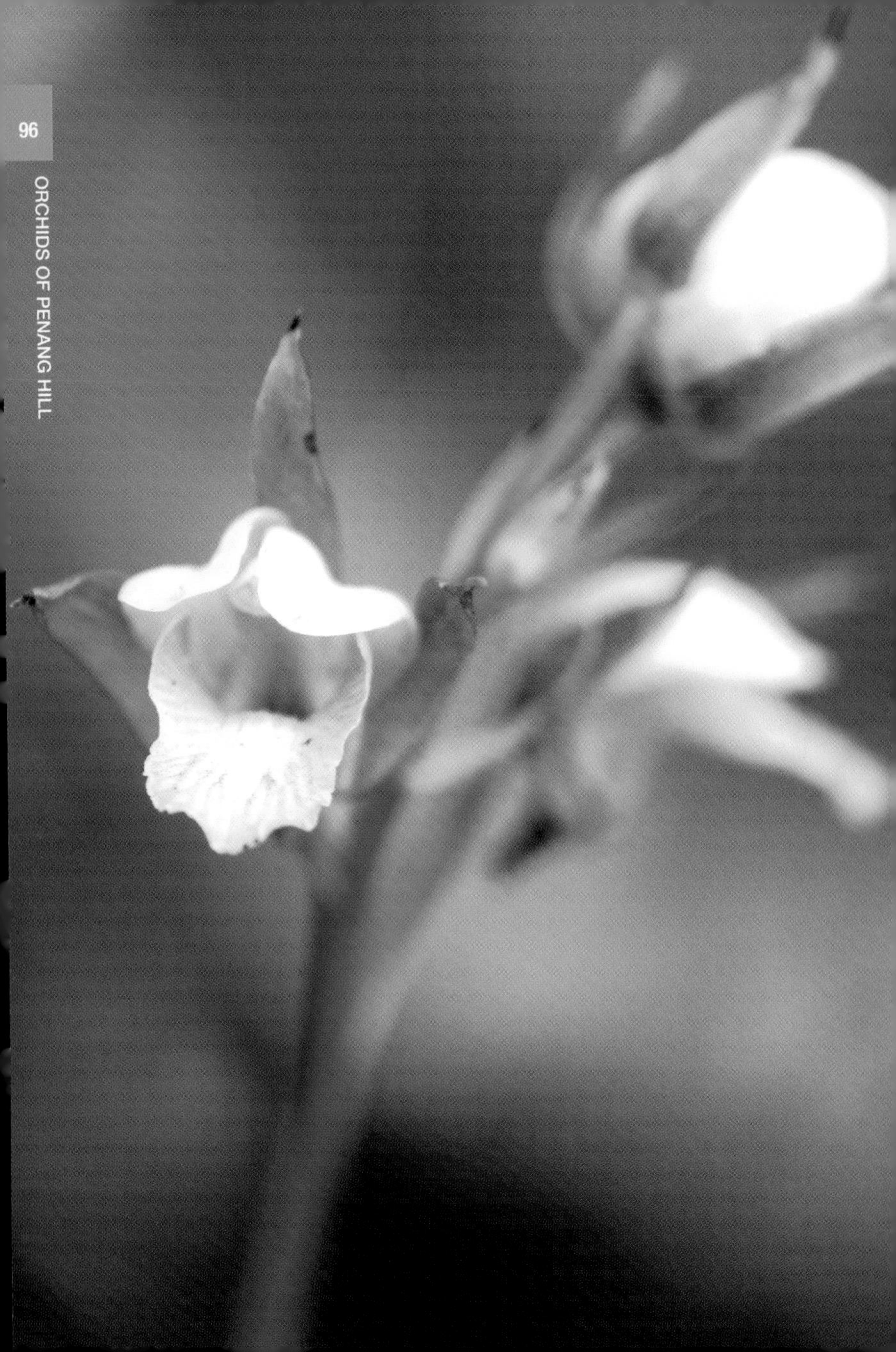

34. *Eulophia nuda* Lindl.

Synonyms:
Cyrtopera fusca Wight
Graphorkis nuda (Lindl.) Kuntze

General Distribution:
India, Myanmar, Thailand,
Peninsular Malaysia, Sumatra,
Java, Borneo and Sulawesi.

Description:

Habitat and ecology:	A terrestrial widely distributed from lowlands to hilly areas and can often be spotted in open areas like grasslands or road cuttings.
Habit:	It has round pseudobulbs about 3 cm in diameter usually partially buried below ground.
Leaves:	Pleated, growing to 15 cm long. Each pseudobulb has 3–4 leaves.
Inflorescence:	Arising from the base of the pseudobulb, growing to 100 cm long. Carries 2–20 flowers.
Flower:	Not opening widely, mildly fragrant. The sepals are narrowly lanceolate-triangular, pale green, flushed brown on the outer surface and are acute. The broadly and obliquely triangular petals are white, sometimes tinged pinkish-purple at the tip. They are curved over the column and lip. The lip is pale mauve with fine pink veins. It has a cone-shaped nectary and is bent in the middle.

In India, the tubers are traditionally used to treat bronchitis, tumours and even snake bites.[41]

35. *Grammatophyllum speciosum* Blume

Common Name:
Tiger Orchid

Synonyms:
Grammatophyllum fastuosum Lindl.
Pattonia macrantha Wight

General Distribution:
Thailand, Peninsular Malaysia,
Sumatra, Java, Borneo, New Guinea
and Solomon Islands.

Description:

Habitat and ecology:	An epiphyte or a lithophyte distributed widely from lowlands to hilly areas. It usually grows near streams, in a ring formation around the trunk of large forest or Durian trees.
Habit:	The pseudobulb looks like a sugarcane stem and is large with many ridges. It can grow to 300 cm long.
Leaves:	Narrow and strap-shaped, curved towards the tip.
Inflorescence:	Arising from the base of the pseudobulb, growing 200 cm long. It is erect to arching, carrying about 40 flowers, which are distant at the lower part but successively closer higher up. Large plants can produce dozens of spikes with a dazzling display of hundreds to thousands of flowers.
Flower:	Large, about 10 cm wide. Sepals and petals are yellow with irregular brown blotches. They are broadly elliptic with the sepals slightly longer than the petals. The trilobed lip has a darker shade of yellow with brown veins, the apex is flushed purple-brown. It has three ridges in the middle from base to near the apex. The column is yellow, spotted purple-brown on the upper surface.
Flowering:	July–September.

It is undoubtedly the heaviest orchid in the world. The *American Orchid Society Bulletin* of 1947 reported a Tiger Orchid specimen with more than 10,000 flowers.[42] While an epiphyte in nature, the tiger orchid is usually grown in the ground (as a terrestrial) because of its enormous size. The seed pods can grow as big as a tropical starfruit.[43]

Grammatophyllum speciosum Blume

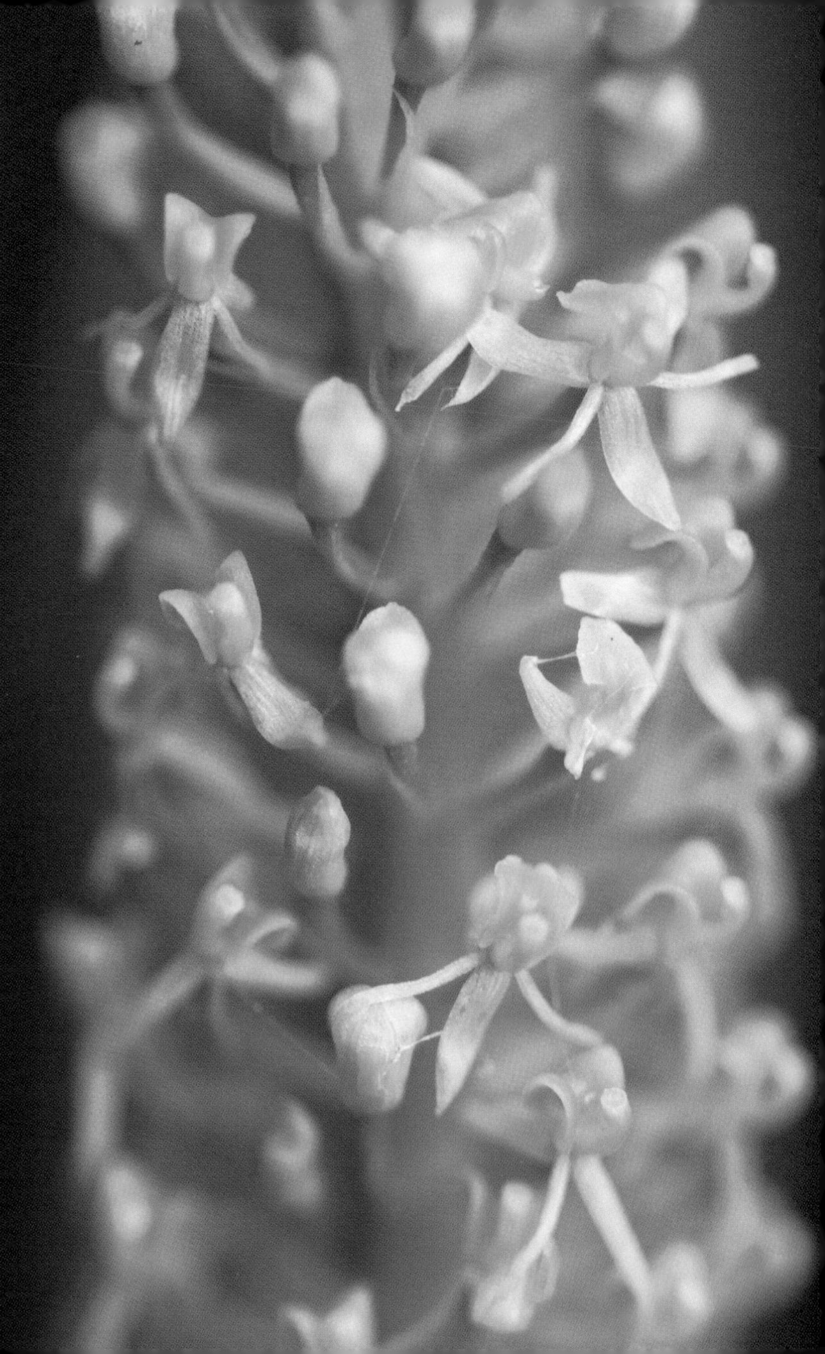

36. *Liparis viridiflora* (Blume) Lindl.

Synonyms:
Malaxis viridiflora Blume
Liparis boothii Regel

General Distribution:
India, Bangladesh, Myanmar, Thailand,
Peninsular Malaysia, Sumatra
and the Philippines.

Description:

Habitat and ecology:	An epiphyte or a lithophyte found in hilly areas.
Habit:	It has narrowly conical pseudobulbs that are laterally flattened. They grow 3–9 cm long.
Leaves:	Oblong-obovate, growing to 27 cm long. Each pseudobulb has 2 leaves.
Inflorescence:	Terminal, growing 15–30 cm long. It is arching, slightly angled and crowded with numerous flowers.
Flower:	Wide opening, small and non-resupinate. The sepals and petals are greenish-white, spreading to reflexed. Dorsal sepal narrowly ovate, apex obtuse, lateral margins revolute. Lateral sepals obliquely and narrowly elliptic, apex obtuse, lateral margins revolute. Petals linear, slightly falcate, apex obtuse, lateral margins sometimes revolute. The lip is recurved in the middle and positioned at the top of the flower. The lip and the column are pale green.

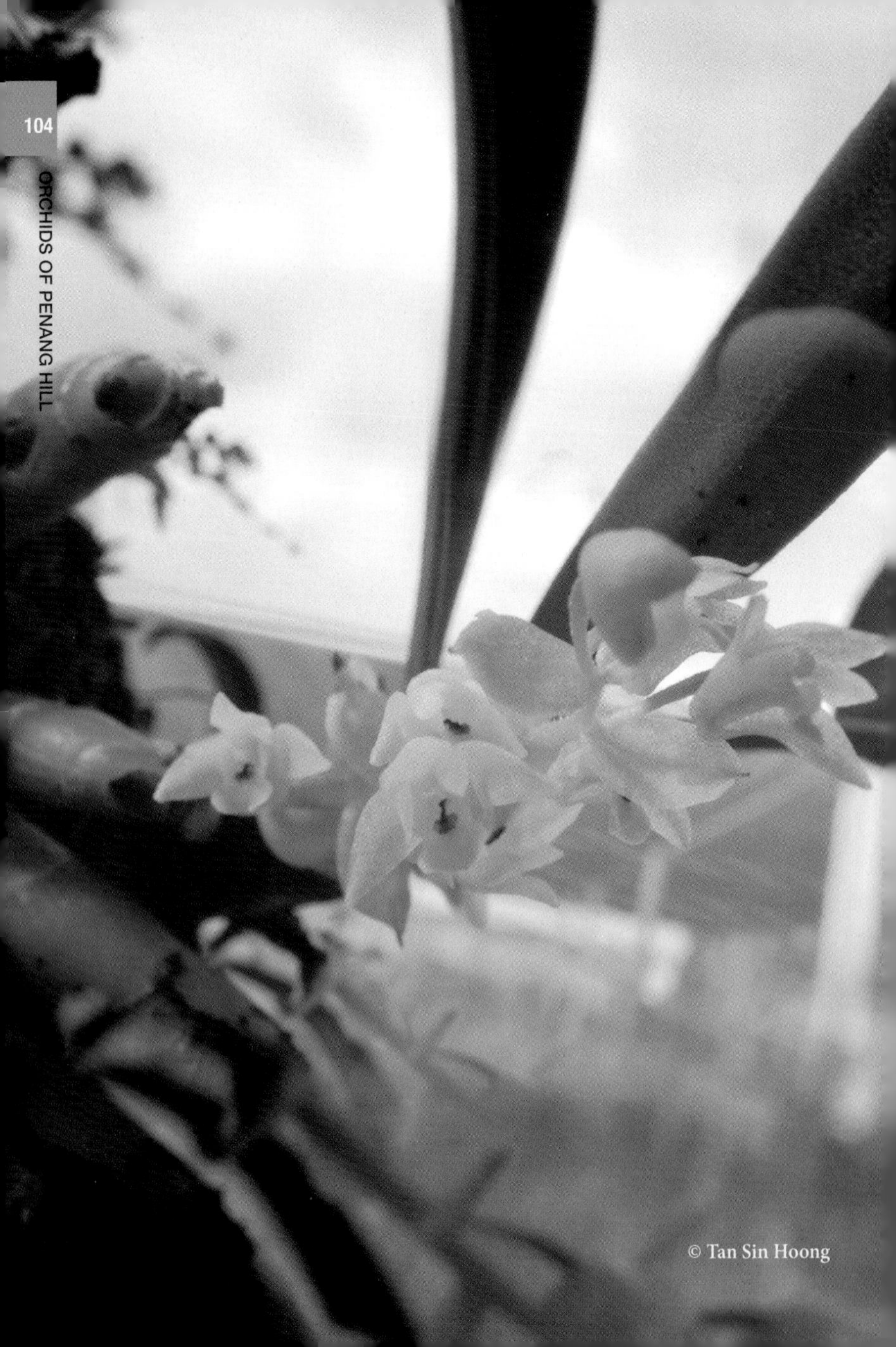

37. *Pinalia maingayi* (Hook.f.) Kuntze

Synonyms:
Eria maingayi Hook.f.

General Distribution:
Endemic to Peninsular Malaysia.

Description:

Habitat and ecology:	An epiphyte found in hilly areas.
Habit:	It has thick and fleshy pseudobulbs about 3 cm long.
Leaves:	Thick, about 8 cm long. Each pseudobulb holds 2–3 leaves.
Inflorescence:	Originates from the apex of the pseudobulb, producing 6–8 flowers.
Flower:	Not wide opening. The sepals and petals are white. The dorsal sepal is ovate, the lateral sepals are obliquely ovate. The petals are elliptic and slightly smaller than the sepals. The lip is white at the basal half, light green at the apical half. The disc has 3 keels: the median keel is purple, at the apical portion of the median keel is a purple blotch, lateral keels cream-pale yellow. The midlobe of the lip is curved downwards near the apex.

38. *Pinalia tenuiflora* (Ridl.) J.J.Wood

Synonyms:
Eria tenuiflora Ridl.
Eria gibbsiae Rolfe

General Distribution:
Myanmar, Thailand, Peninsular Malaysia, Sumatra, Java and Borneo.

Description:

Habitat and ecology:	A widespread epiphyte which usually grows on tree branches in the lowlands to the hilly areas.
Habit:	It has erect stems which can grow to 20 cm long.
Leaves:	Thin, oblong-elliptic, growing to 16 cm long. Each stem has 3–5 leaves.
Inflorescence:	Originates from axils below the leaves, growing 6 cm long. It spreads horizontally, holding 20–40 flowers.
Flower:	Sepals and petals are yellow and narrowly ovate-elliptic. Petals are smaller than the sepals. The lip is yellow at its tip with a deep purple patch at its base. The lip has slightly raised sides with an acute tip.
Flowering:	December–January.

39. *Plocoglottis javanica* Blume

Common Name:
Mousetrap Orchid

Synonym:
Plocoglottis fimbriata Teijsm. & Binn.

General Distribution:
Andaman Islands, Thailand,
Peninsular Malaysia, Sumatra,
Java and Borneo.

Description:

Habitat and ecology:	A terrestrial which grows in relatively moist and dark areas of the forest in hilly areas.
Habit:	The plant has a cylindrical, slender pseudobulb about 10 cm long.
Leaves:	Elliptical, with a long stalk. Each pseudobulb has one leaf.
Inflorescence:	Arising from the base of the pseudobulb, growing 90 cm long. It is erect, unbranched, holding numerous well-spaced flowers which are short-lived.
Flower:	Not wide opening, about 1.7 cm across. The sepals and petals are yellow with irregular dark red markings. The sepals are oblong with acute tips while the petals are much narrower. The lip is yellow, paler at the base and is oblong-ovate with a caudate apex that is curved downwards and inwards, usually obscured from view. The column is yellow, marked purple-red ventrally; the base is cream-white ventrally.
Flowering:	September–October.

The lip of this species has a unique 'snapping' movement. Much like a mousetrap, the lip springs upward near the flower column when an insect touches the lip. This action is believed to improve the chances of pollination as the insect is temporarily ensnared and, in the process of freeing itself, the insect carries the pollen as it manoeuvres out of the flower.[44]

40. *Podochilus microphyllus* Lindl.

Synonym:
Podochilus confusus J.J.Sm.

General Distribution:
Myanmar, Vietnam, Cambodia, Thailand, Peninsular Malaysia, Sumatra, Java and Borneo.

Description:

Habitat and ecology:	An epiphyte or a lithophyte which usually grows on large granitic boulders close to streams in hilly areas.
Habit:	The small pendulous stems grow to 20 cm long.
Leaves:	Twisted at the base with a pointed tip. They are arranged alternately on the stem.
Inflorescence:	Terminal. It is pendulous and carries 2–6 flowers.
Flower:	About 4 mm across. The sepals are white, sometimes marked purple in the median. The dorsal sepal is ovate, apex acute. The lateral sepals are joined at inner margins to form a mentum. The petals are slightly narrower than the sepals, white with a purple marking in the median. The lip is white and trowel-shaped. A greenish-white form without purple markings on the sepals and petals has been sighted in Penang.

Podochilus microphyllus Lindl.

41. *Renanthera elongata* (Blume) Lindl.

Common Name:
Pokok Api Sesudah

Synonyms:
Aerides elongata Blume
Renanthera micrantha Blume

General Distribution:
Thailand, Peninsular Malaysia, Sumatra, Java, Borneo and the Philippines.

Description:

Habitat and ecology:	An epiphyte. The sun-loving plant is widely distributed from lowland to hilly areas and grows in relatively open areas, clambering over tree crowns. It creeps up host trees as a climber with adventitious roots which grow from its sides.
Habit:	It has very long stems with well-spaced internodes.
Leaves:	Narrowly oblong and arranged alternately on the stems.
Inflorescence:	Arising laterally and horizontally on the stem between the leaves. It usually has many branches carrying numerous flowers.
Flower:	Wide opening, almost entirely red. The sepals and petals are pale orange-red with darker red markings, sometimes flushed entirely red. The dorsal sepal is oblong-obtuse while the lateral sepals are similar but broader. Petals are spatulate, obtuse, narrower than the sepals. The small, trilobed lip is similarly coloured but has a white blotch in the middle. Its midlobe is recurved. The column is red.

The plant loves sunlight and has been observed to only begin flowering once it grows over the tree crown and is exposed to full sunlight.[45] The vernacular Malay name 'Pokok Api Sesudah' translates to 'Tree After Fire'.[46] This is partly due to the nature of the plant which climbs over trees and sends out numerous spikes that flower together in season, covering the crown of the tree in a spectacular display of red flowers.

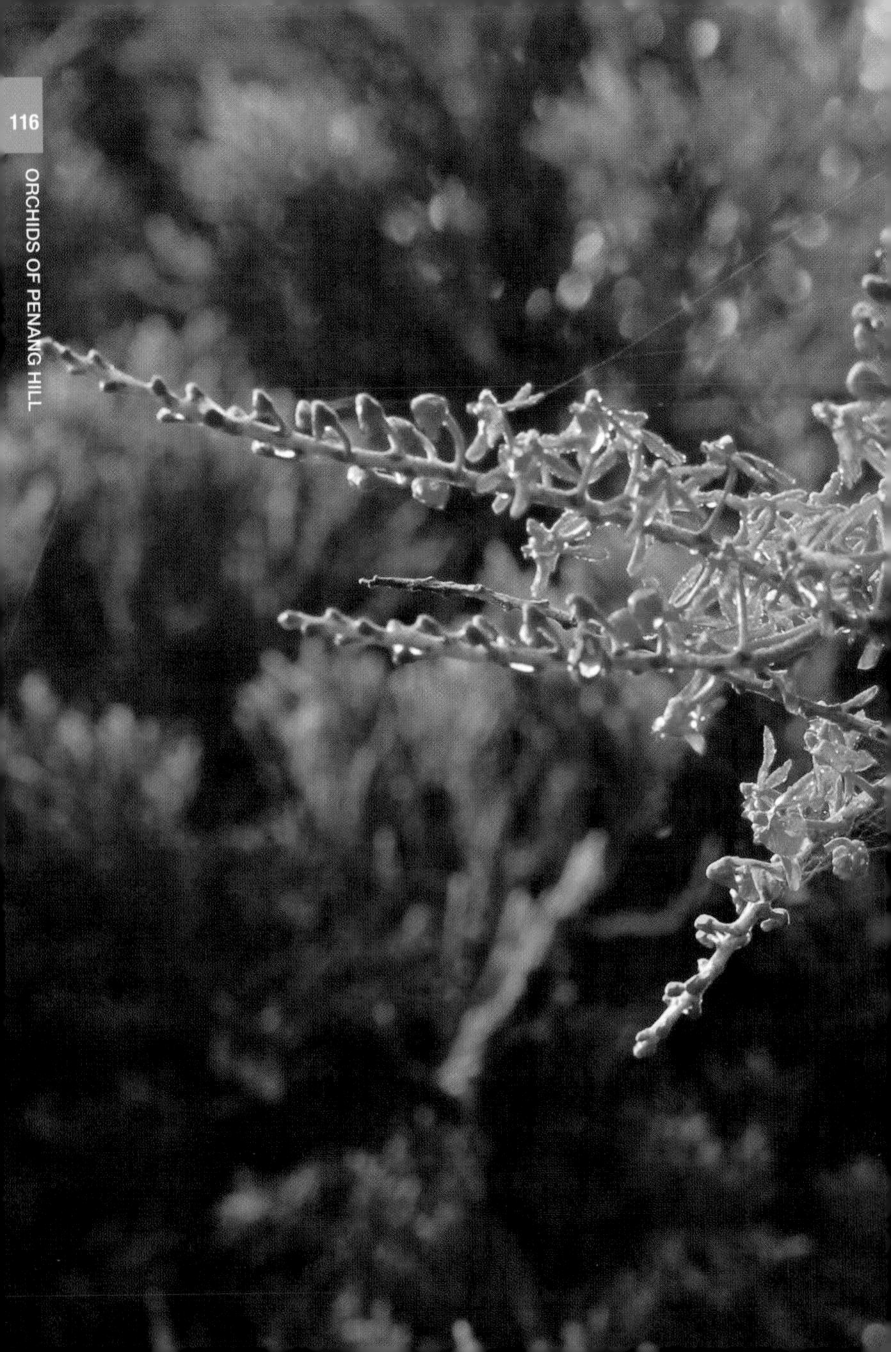

Renanthera elongata (Blume) Lindl.

42. *Spathoglottis plicata* Blume

Common Name:
Orkid Pinang

Synonyms:
Calanthe poilanei Gagnep.
Phaius rumphii Blume

General Distribution:
India, Myanmar, Thailand,
Peninsular Malaysia and Pacific Islands.

Description:

Habitat and ecology:	A terrestrial or a lithophyte, widely distributed from lowland to hilly areas, often spotted in secondary growth, road cuttings and former quarry sites.
Habit:	It has small crowded pseudobulbs growing to 50 mm long.
Leaves:	Pleated and palm-like, arranged around a rigid stalk that is firmly rooted. Each pseudobulb has 3–4 leaves.
Inflorescence:	Arising from the base of the pseudobulb, growing 50–100 cm long, bearing many successive flowers.
Flower:	About 4 cm across. The sepals and petals are pinkish purple. The sepals are ovate and the petals are broadly ovate-elliptic. The lip is long and slender, broadening at the mid-section, the apex is bilobed, sometimes with an apicule in the median. The lip's midlobe is purple at the apical half, yellow at the basal half, marked purple at the base, the side lobes are purple, the disc is yellow with two large yellow calli. Flowers tend to be self-pollinating and form seed pods easily.
Flowering:	Throughout the year.

A very common species in hilly areas of Penang. It is also a popular house plant and many attractive hybrids have been bred from it. The rare pure white form is called 'Penang White'.

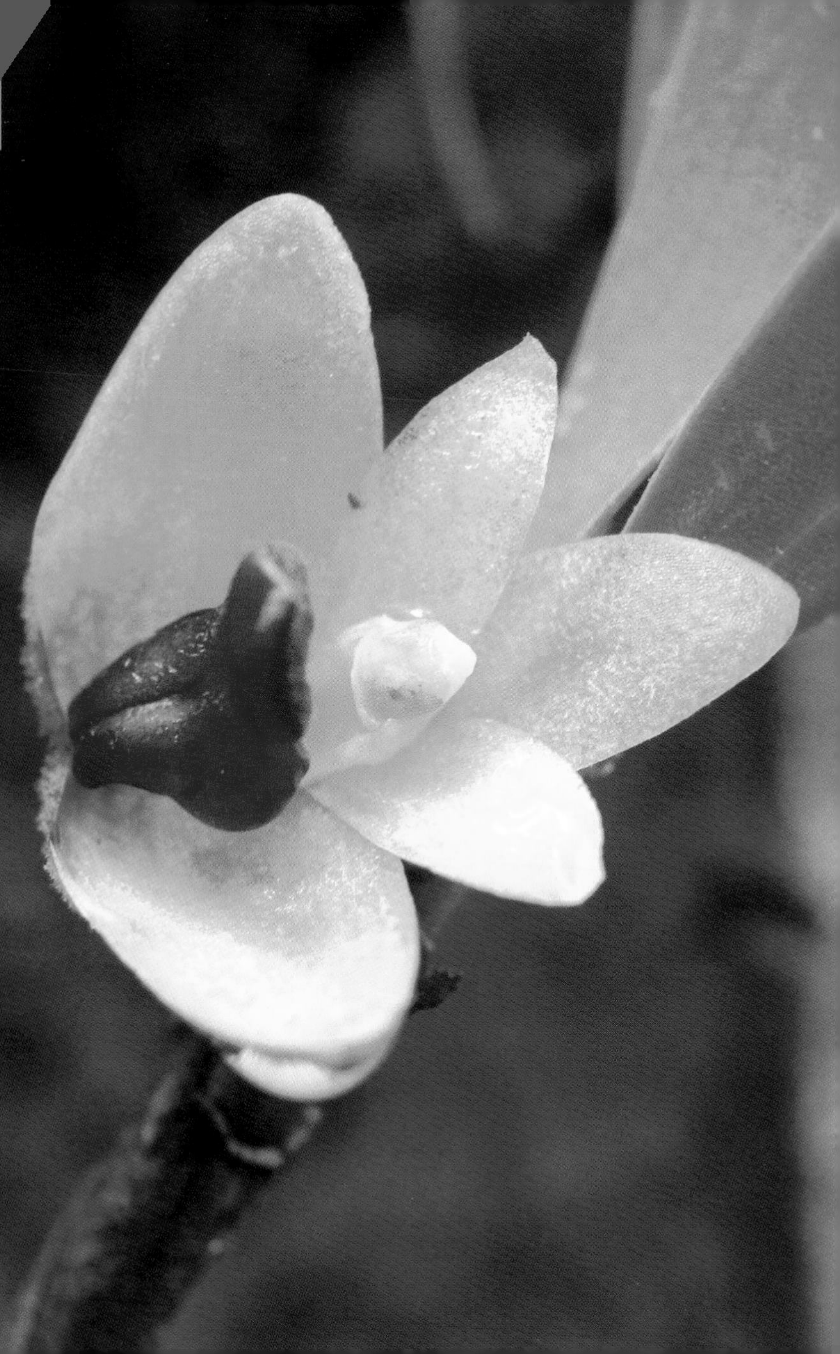

43. *Strongyleria leiophylla* (Lindl.) Schuit., Y.P.Ng & H.A.Pedersen

Synonyms:

Campanulorchis leiophylla (Lindl.) Y.P.Ng & P.J.Cribb

Eria leiophylla Lindl.

General Distribution:

Peninsular Malaysia, Sumatra, Borneo, Sulawesi and Maluku.

Description:

Habitat and ecology:	An epiphyte which grows on tree trunks and branches in hilly areas.
Habit:	The plant has stout rhizomes with erect pseudobulbs. It tends to spread quickly to form large clumps. The pseudobulb grows to 9 cm long, and tapers to a slender apex.
Leaves:	Narrowly oblong-elliptic, to about 15 cm long. Each shoot bears 2–3 leaves.
Inflorescence:	Arising near the top of the pseudobulb. It is short, covered in red-brown hairs and carries 1–2 flowers.
Flower:	Wide opening and non-resupinate. The sepals and petals are pale to dark yellow on the abaxial, the adaxial is covered with pale brown hairs. The dorsal sepal is ovate-elliptic. The lateral sepals are obliquely ovate-triangular, much broader than the dorsal sepal. The petals are narrowly oblong-elliptic. The lip is fleshy and dark maroon, the callus on the lip is dark maroon-black. Column and anther are yellow.

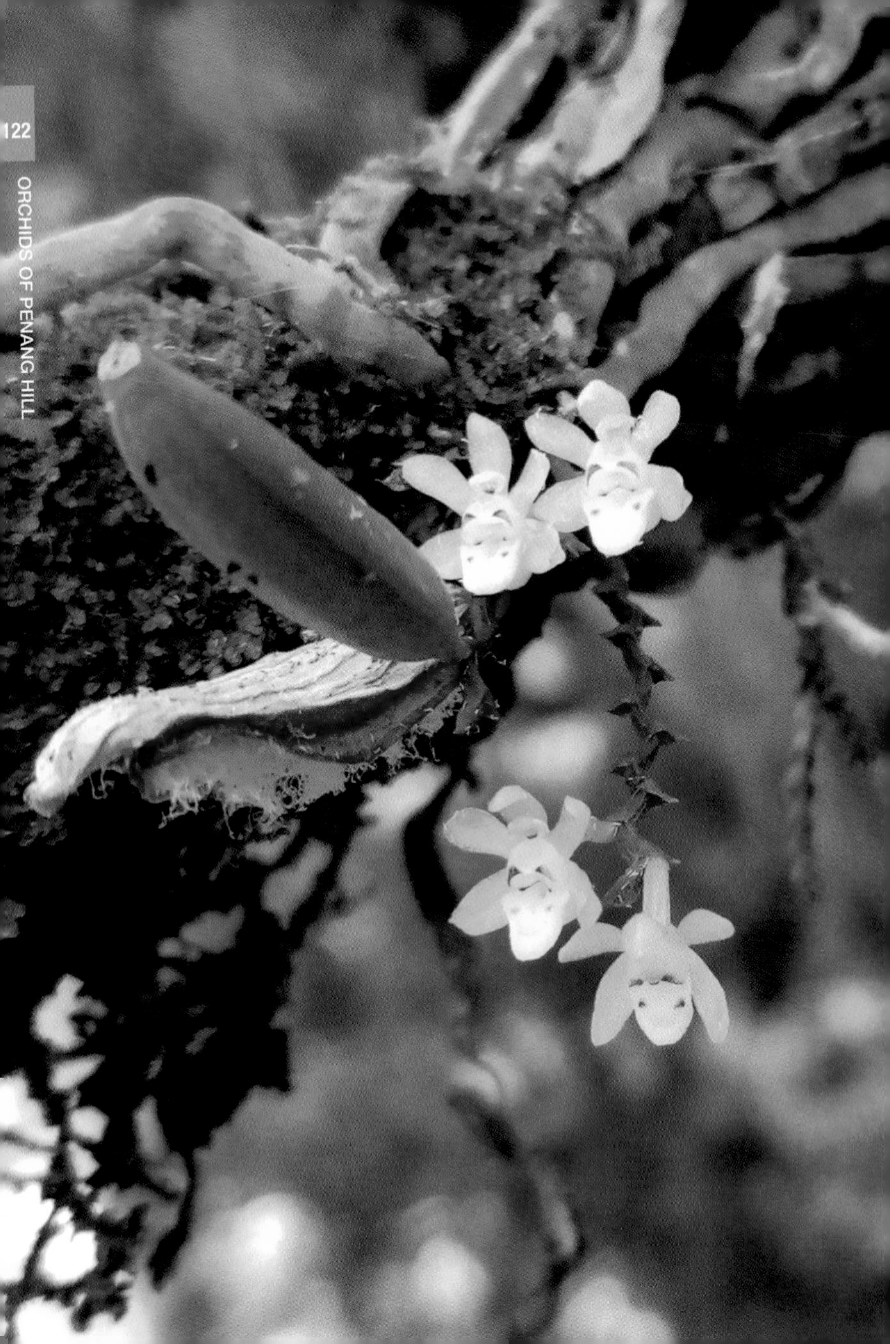

44. *Taeniophyllum hasseltii* Rchb.f.

Common Name:
Ribbon Root Orchid

Synonym:
Taeniophyllum calceolus Carr

General Distribution:
Peninsular Malaysia, Java and Christmas Island.

Description:

Habitat and ecology:	An epiphyte found in the lowlands on twigs and occasionally tree trunks.
Roots:	Greyish-green, stout and fleshy. Roots are photosynthetic.
Habit:	It has inconspicuous stems which grow to 2 mm long.
Leaves:	It has small triangular leaf scales.[47]
Inflorescence:	Arising from the axil of the lower stem. It is held on a 4–6 mm long peduncle with flowers arising successively, bearing up to 10 flowers. 1–2 flowers bloom at a time.
Flower:	Wide opening and spurred at the base. The sepals and petals are pale to dark yellow. The dorsal sepal is elliptic while the lateral sepals are obliquely ovate-elliptic. The petals are narrower than the sepals, oblong and slightly curved. The sepals and petals are curved forward. The trilobed lip is cream-white with two purple blotches. It is spurred, and projects forward with its hook-like apex curved upwards. The column is cream-pale yellow, margins of the stigma purple and the anther is pale yellow. Flowers only last one day.
Flowering:	May–July.

As this species grows on twigs, it is ever susceptible to falling to the ground. Hence it grows, matures and flowers fast, ensuring the continuity of the species.

45. *Tainia speciosa* Blume

Synonyms:
Eria speciosa (Blume) Rchb.f.
Mitopetalum speciosum (Blume) Blume

General Distribution:
Thailand, Peninsular Malaysia, Sumatra, Java and Borneo.

Description:

Habitat and ecology:	A terrestrial distributed in hilly areas, growing on the shaded forest floor with deep humus.
Habit:	It has pseudobulbs that grow 6 cm long.
Leaves:	Plicate, growing to 15 cm long. Each pseudobulb supports a single leaf.
Inflorescence:	Arising from the base of the pseudobulb, growing to 40 cm long. It carries 4–12 flowers.
Flower:	Wide opening. Sepals and petals are greenish-yellow with dark purple veins. They are wide at the base and taper to a slender tip. The lip is white to pale yellow and has a hairy midline. Its midlobe is curved downwards with a pointed tip. The column is pale yellow.
Flowering:	May.

46. *Thelasis carinata* Blume

Common Name:
Triangular Fly Orchid

Synonyms:
Eria carinata (Blume) Rchb.f.
Oxyanthera carinata (Blume) Schltr.

General Distribution:
Thailand, Peninsular Malaysia, Sumatra,
Java, Borneo, the Philippines, Sulawesi
and Maluku.

Description:

Habitat and ecology:	An epiphyte or a lithophyte distributed in hilly areas, grows on large tree branches.
Habit:	It has pale green, fleshy stems which are short and tufted.
Leaves:	Narrow and oblong. Each stem supports about 5 leaves.
Inflorescence:	Arising from the leaf axils. The thin arching scape grows to about 25 cm long, supporting up to 20 flowers.
Flower:	Not opening widely, small. Sepals and petals form a short tube at the base. The sepals and petals are pale white with a greenish base. Dorsal sepal concave at base half, reflexed at apical half, narrowly triangular in outline, apex attenuate to acuminate. Lateral sepals gently curved, concave at base, apex attenuate to acuminate. Petals spreading laterally at apex, narrowly triangular, apex attenuate to acuminate. Lip is white, with a concave base which has a short, pointed tip and a nectary on each side.
Flowering:	June–August.

47. *Thrixspermum annamense* (Guillaumin) Garay

Synonyms:
Thrixspermum austrosinense Tang & F.T.Wang
Thrixspermum devolium T.P.Lin & C.C.Hsu

General Distribution:
China, Taiwan, Vietnam, Thailand
and Peninsular Malaysia.

Description:

Habitat and ecology:	An epiphyte found in hilly areas growing on the branches and stems of shrubs.
Habit:	It has a short, erect stem.
Leaves:	Narrowly oblong and stiff, measuring about 3 cm long.[48]
Inflorescence:	Arising from the leaf axils on the stem, growing 5–7 cm long, carrying successive flowers; 1–2 flowers bloom at a time.
Flower:	5 mm across, mildly fragrant. The sepals and petals are white. The sepals are ovate and larger than the petals. The lip is pale yellow with faint brown markings. It has an upturned midlobe with a tuft of short club-shaped hairs near the tip. The column and anther are white. Flowers only last one day.

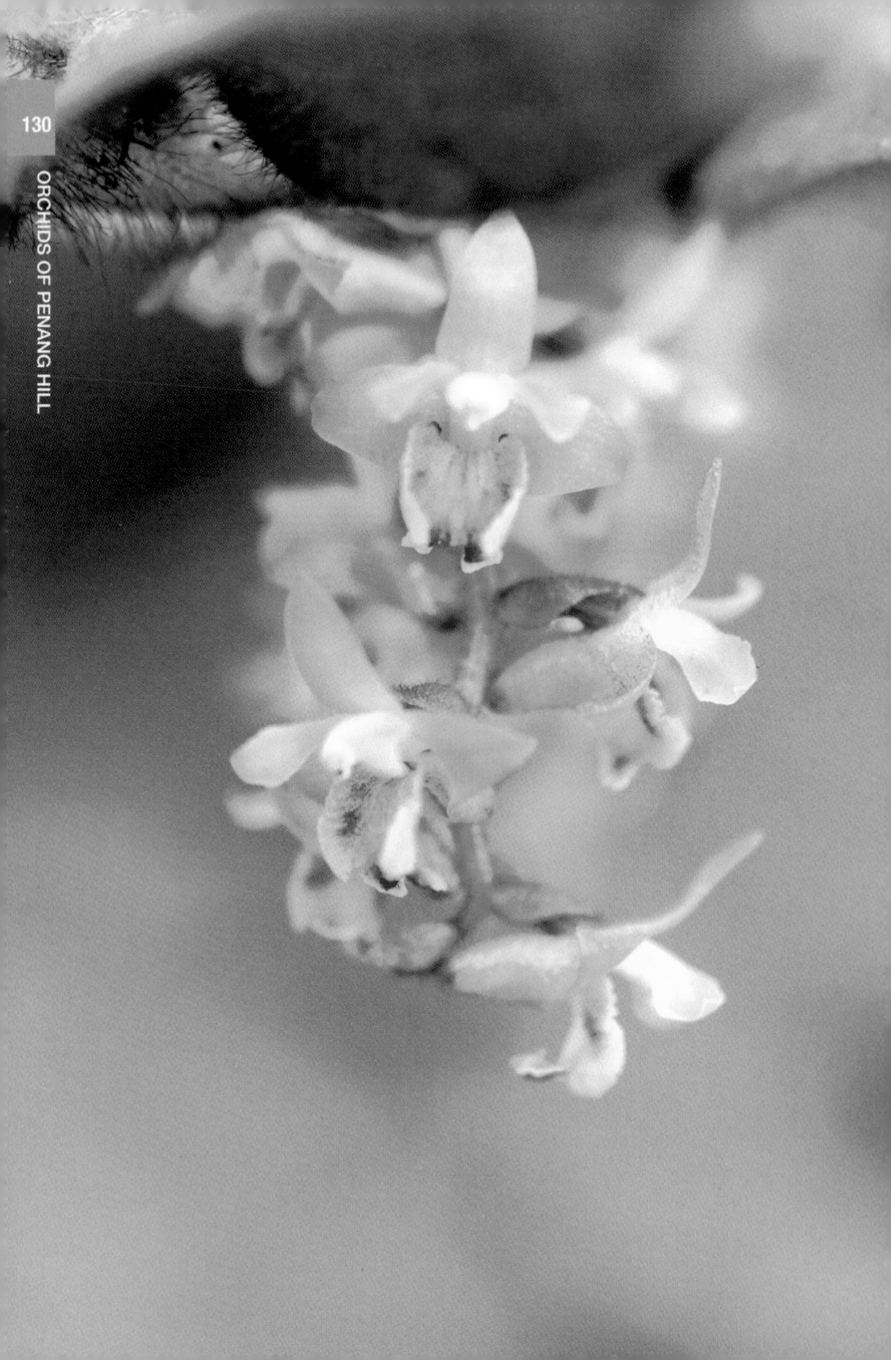

48. *Trichotosia ferox* Blume

Synonyms:
Eria ferox (Blume) Blume
Pinalia ferox (Blume) Kuntze

General Distribution:
Thailand, Peninsular Malaysia, Sumatra, Java, Borneo and Lombok.

Description:

Habitat and ecology:	An epiphyte primarily found in hilly areas.
Habit:	It has long pendulous stems covered with fine red-brown hairs.
Leaves:	Oblong, covered in fine red-brown hairs, arranged alternately on the stems.
Inflorescence:	Emerging from between the internodes on the stem, growing 5–10 cm long. It is pendulous, pilose, supporting up to 10 flowers. Many stalks arise between the internodes on the same stem.
Flower:	Spreading, greenish-yellow. Dorsal sepal suberect, apex slightly reflexed, oblong-ovate, apex subacute to acute. Lateral sepals obliquely oblong-ovate, apex acute. Sepals are hairy on the adaxial. Petals slightly obliquely oblong, apex obtuse-rounded, glabrous. The trilobed lip has a greenish-yellow base with light red blotches, sometimes entirely flushed purple-red. The midlobe of the lip is curved downwards, apex notched, the disc with 3–5 rows of hairy, warty keels. Column is white to greenish-yellow and anther is cream-white, flushed pale purple.
Flowering:	July–September.

ORCHIDOIDEAE

49. *Anoectochilus geniculatus* Ridl.

Common Name:
Jewel Orchid

General Distribution:
Myanmar, Thailand, Peninsular Malaysia, Sumatra and Borneo.

Description:

Habitat and ecology:	A terrestrial usually found in hilly areas, growing in the moist humus of the forest floor, preferring areas of low light and high humidity.
Habit:	It has a short creeping rhizome.
Leaves:	Ovoid, dark purple with contrasting copper-red veins. Each plant has about 4 leaves.
Inflorescence:	Terminal, growing to 14 cm long. The stalk is erect, holding a succession of 2–5 flowers.
Flower:	The sepals are white on the abaxial, flushed purple-red on the adaxial. They are hairy on the adaxial. The dorsal sepal and petals form a hood. The lateral sepals spreading, a little shorter than the dorsal sepal. The petals are white on both surfaces, glabrous. The lip is white, with a shortly pointed pale yellow spur, sometimes slightly translucent. The claw is strongly bent to the base of the lip with thin filaments on each side, the end recurved, bearing a 2-lobed blade.

Prized for its attractive foliage, it has fallen prey to orchid poaching, becoming rare throughout its distribution range. Collected plants sometimes end up in terrariums as they grow in conditions of low light and high humidity.

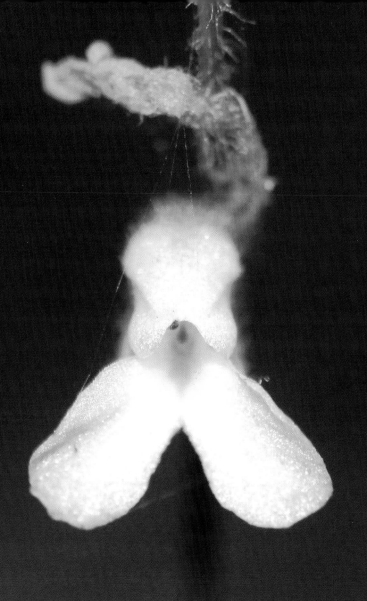

50. *Zeuxine rupestris* Ridl.

General Distribution:
Endemic to Peninsular Malaysia.

Description:

Habitat and ecology:	A terrestrial found in hilly areas where it grows in leaf litter.
Habit:	It has a slender, hairy stem. The plant grows 12–20 cm tall.
Leaves:	Oval, 1.5–2 cm long. They are in the basal half of the stem.
Inflorescence:	Terminal, about 8 cm long. It is erect and hairy, usually carrying 2 flowers.
Flower:	Entirely white. Sepals hairy on the adaxial. The dorsal sepal and petals form a hood. The lip is the most prominent part of the flower, with a narrow neck leading to a blade of two oblong, slightly diverging lobes at a right angle to the axis of the lip.
Flowering:	December–January.

ENDNOTES

1. Ong *et al.*, *Wild Orchids of Peninsular Malaysia,* 2011, p. 7.
2. Fadelah *et al.*, *Orchids: The Living Jewels of Malaysia*, 2001, p. 14.
3. Smith, 'William John Swainson: who was he?', 2013.
4. Harris, 'Orchidelirium: The obsession with orchids', 2019.
5. Royal Botanic Gardens, Kew, 'World Checklist of Selected Plant Families', 2020.
6. WWF Malaysia, Study on the development of hill stations, 2001, p. 53.
7. Rusea *et al.*, 'An assessment of orchids' diversity in Penang Hill, Penang, Malaysia after 115 years', 2011, p. 2264.
8. Curtis and Ridley, 'A catalogue of the flowering plants and ferns found growing wild in the island of Penang', 1894.
9. Rusea *et al.*, 'An assessment of orchids' diversity in Penang Hill, Penang, Malaysia after 115 years', 2011.
10. Rusea and Nordin, 'Orchids of Penang Hill, an updated checklist', 2017.
11. Nga *et al.*, 'Five new records of terrestrial and lithophytic orchids (Orchidaceae) from Penang Hill, Malaysia', 2016.
12. Straits Settlements, Government Gazette, 1891.
13. Teoh, *Orchids of Asia*, 2005, pp. 96, 102.
14. Sim, 'Flowers of Penang Hill', 1950.
15. Rusea et al., 'An assessment of orchids' diversity in Penang Hill, Penang, Malaysia after 115 years', 2011, p. 2263.
16. Ong *et al.*, *Wild Orchids of Peninsular Malaysia*, 2011, p. 183.
17. Ong, 'Flora of Peninsular Malaysia – Apostasioideae', 2013, pp. 98–101.
18. Teoh, *Medicinal Orchids of Asia*, 2016, p. 117.
19. Ong, 'Flora of Peninsular Malaysia – Apostasioideae', 2013, pp. 102–05.
20. Non-Timber Forest Products Centre of Excellence, 'Lesser wildly used medicinal plants', 2015.
21. Fan et al., 'Rain pollination provides reproductive assurance in a deceptive orchid', 2012, pp. 956–57.
22. Teoh, *Medicinal Orchids of Asia*, 2016, p. 88.
23. Nongdam, 'Ethno-medicinal uses of some orchids of Nagaland, North-East India', 2014.
24. Debnath et al., '*Arundina graminifolia* (D.Don) Hochr (Orchidaceae) – a new addition to the Flora of Tripura', 2016.
25. Nongdam, 'Ethno-medicinal uses of some orchids of Nagaland, North-East India', 2014.

26 Teoh, *Medicinal Orchids of Asia*, 2016, p. 122.

27 Singapore National Parks Board, '*Arundina graminifolia* (D.Don) Hochr.', 2019.

28 Teoh, *Medicinal Orchids of Asia*, 2016, p. 147.

29 Singapore National Parks Board, '*Bromheadia finlaysoniana* (Lindl.) Miq.', 2013.

30 Teoh, *Medicinal Orchids of Asia*, 2016, p. 148.

31 *Ibid.*

32 Ong and Tan, 'Fly pollination in four Malaysian species of *Bulbophyllum* (section *Sestochilus*)', 2011, pp. 103–05.

33 Singapore National Parks Board, '*Bulbophyllum medusae* (Lindl.) Rchb.f.', 2016.

34 Singapore National Parks Board, '*Bulbophyllum vaginatum* (Lindl.) Rchb.f.', 2019.

35 Lok *et al.*, 'The status and distribution in Singapore of *Calanthe pulchra* (Bl.) Lindl.', 2010, p. 88.

36 Teoh, *Medicinal Orchids of Asia*, 2016, p. 222.

37 *Ibid.*

38 Teoh, *Medicinal Orchids of Asia*, 2016, p. 271.

39 Singapore National Parks Board. '*Dendrobium crumenatum* Sw.', 2013.

40 Teoh, *Medicinal Orchids of Asia*, 2016, p. 297.

41 Patil and Mahajan, 'Ethnobotanical potential of *Eulophia* species for their possible biological activity', 2013, p. 302.

42 Teo, *Native Orchids of Peninsular Malaysia*, 1985, p. 54.

43 Ong *et al.*, *Wild Orchids of Peninsular Malaysia*, 2011, p. 31.

44 Ang *et al.*, 'The status and distribution in Singapore of *Plocoglottis javanica* Blume (Orchidaceae)', 2011, p. 73.

45 Ang *et al.*, 'Rediscovery of *Renanthera elongata* (Blume) Lindl (Orchidaceae) in Singapore', 2011, pp. 298–300.

46 *Ibid.*

47 Ong *et al.*, 'Clarification of the status of *Taeniophyllum hasseltii* Rchb.f in Peninsular Malaysia', 2019, p. 86.

48 Chen and Wood, '*Thrixspermum annamense*', 2009, p. 467.

REFERENCES

Ang, W.F., Lok, A.F.S.L., Yeo, C.K., Ng, A.A.P.X., Ng, B.Y.Q., and Tan, H.T.W., 'Rediscovery of *Renanthera elongata* (Blume) Lindl. (Orchidaceae) in Singapore', *Nature in Singapore*, 4 (2011): 297–301.

Ang, W.F., Lok, A.F.S.L., Yeo, C.K., Ng, B.Y.Q., and Tan, H.T.W., 'The status and distribution in Singapore of *Plocoglottis javanica* Blume (Orchidaceae)', *Nature in Singapore*, 4 (2011): 73–77.

Chen Xinqi and Gale, S.W., '*Arundina* Blume', *Flora of China*, 25 (2009): 314–15, Beijing: Science Press; St. Louis, Missouri: Botanical Garden Press. Retrieved from http://www.efloras.org/florataxon.aspx?flora_id=2&taxon_id=102739

Chen Xinqi and Wood, J.J., '*Thrixspermum* Loureiro, Fl. Cochinch', *Flora of China*, 25 (2009): 467, Beijing: Science Press; St. Louis, Missouri: Botanical Garden Press. Retrieved from http://www.efloras.org/florataxon.aspx?flora_id=2&taxon_id=242414488

Curtis, C., and Ridley, H. N., 'A catalogue of the flowering plants and ferns found growing wild in the island of Penang', *Journal of the Straits Branch of the Royal Asiatic Society*, 25 (1894): 67–167.

Debnath, B., Sarma, D., Paul, C., and Debnath, A., '*Arundina graminifolia* (D.Don) Hochr (Orchidaceae) – a new addition to the Flora of Tripura', *Asian Journal of Plant Science and Research*, 6(3) (2016): 28–31.

Fadelah Abdul Aziz, Zaharah Hasan, Rozlaily Zainol, Nuraini Ibrahim, Tan Swee Lian and Hamidah Sulaiman, *Orchids: The Living Jewels of Malaysia*, Kuala Lumpur: Malaysian Agricultural Research and Development Institute (MARDI), 2001.

Fan, X., Barrett, S.C.H., Lin, H., Lingling, C., Zhou, X., and Gao, J. Y., 'Rain pollination provides reproductive assurance in a deceptive orchid', *Annals of Botany*, 110 (2012): 953–58.

Harris, B.R., 'Orchidelirium: The obsession with orchids', *History Daily*, 18 January 2019. Retrieved from https://historydaily.org/orchidelirium-the-obsession-with-orchids

Lok, A.F.S.L., Ang, W.F., and Tan, H.T.W., 'The status and distribution in Singapore of *Calanthe pulchra* (Bl.) Lindl.', *Nature in Singapore*, 3 (2010): 87–90.

Nga, S.Y., Nordin, F.A., and Othman, A.S., 'Five new records of terrestrial and lithophytic orchids (Orchidaceae) from Penang Hill, Malaysia', *Tropical Life Sciences Research*, 27(2) (2016): 103–09.

Non-Timber Forest Products Centre of Excellence, 'Lesser widely used medicinal plants', 2015. Retrieved from http://nce.gov.in/listofNTFP/NTFPMEDICINAL.pdf

Nongdam, P., 'Ethno-medicinal uses of some orchids of Nagaland, North-East India', *Research Journal of Medicinal Plants*, 8 (2014): 126–39. https://scialert.net/abstract/?doi=rjmp.2014.126.139.

Ong, P.T., 'Flora of Peninsular Malaysia – Apostasioideae', *Malesian Orchid Journal*, 12 (2013): 93–116.

Ong, P.T. and Tan, K.H., 'Fly pollination in four Malaysian species of *Bulbophyllum* (section *Sestochilus*) – *B. lasianthum, B. lobbii, B. subumbellatum* and *B. virescens*', *Malesian Orchid Journal*, 8 (2011): 103–10.

Ong, P.T., O'Byrne, P., Saw, L.G., and Chung, R.C.K., 'Checklist of Orchids of Peninsular Malaysia', Kepong: Forest Research Institute Malaysia, Research Pamphlet No. 136, 2017.

Ong, P.T., O'Byrne, P., Yong, W.S.Y., and Saw, L.G., *Wild Orchids of Peninsular Malaysia*, Kepong: Forest Research Institute Malaysia, 2011.

Ong, P.T., Tan, J.P.C., and Chacko, R.P., 'Clarification of the status of *Taeniophyllum hasseltii* Rchb.f. in Peninsular Malaysia', *Malayan Orchid Review*, 53 (2019): 85–89.

Patil, M.C. and Mahajan, R.T., 'Ethnobotanical potential of *Eulophia* species for their possible biological activity', *International Journal of Pharmaceutical Sciences Review and Research*, 21(2) (2013): 297–307.

Rusea Go and Nordin, F.A., 'Orchids of Penang Hill, an updated checklist', *Folia Malaysiana*, 18(1) (2017): 21–50.

Rusea Go, Eng, K.H., Mustafa, M., Abdullah, J.O., Naruddin, A.A., Lee, N.S., Lee, C.S., Eum, S.M., Park, K.W., and Choi, K., 'An assessment of orchids' diversity in Penang Hill, Penang, Malaysia after 115 years', *Biodiversity and Conservation*, 20(10) (2011): 2263–72. https://doi.org/10.1007/s10531-011-0087-z,.

Sandrasagaran, U.M., Subramaniam, S., and Murugaiyah, V., 'New perspective of *Dendrobium crumenatum* orchid for antimicrobial activity against selected pathogenic bacteria', *Pakistan Journal of Botany*, 46(2) (2014): 719–24.

Seidenfaden, G. and Wood, J.J., *The Orchids of Peninsular Malaysia and Singapore*, Fredensborg, Denmark: Olsen-Olsen, 1992.

Sim, K., 'Flowers of Penang Hill', *The Singapore Free Press*, 20 May 1950, Retrieved from https://eresources.nlb.gov.sg/newspapers/Digitised/Article/freepress19500520-1.2.139

Singapore National Parks Board, '*Arundina graminifolia* (D.Don) Hochr.', NParks Flora & Fauna Web, 2019. Retrieved 3 September 2019 from https://www.nparks.gov.sg/florafaunaweb/flora/1/6/1690

Singapore National Parks Board, '*Bromheadia finlaysoniana* (Lindl.) Miq.', NParks Flora & Fauna Web, 2013. Retrieved 23 December 2018 from https://www.nparks.gov.sg/florafaunaweb/flora/3/7/3742

Singapore National Parks Board, '*Bulbophyllum medusae* (Lindl.) Rchb.f.', NParks Flora & Fauna Web, 2016. Retrieved 3 September 2019 from https://www.nparks.gov.sg/florafaunaweb/flora/4/9/4955

Singapore National Parks Board, '*Bulbophyllum vaginatum* (Lindl.) Rchb.f.', NParks Flora & Fauna Web, 2019. Retrieved 3 September 2019 from https://www.nparks.gov.sg/florafaunaweb/flora/3/6/3648

Singapore National Parks Board, '*Dendrobium crumenatum* Sw.', NParks Flora & Fauna Web, 2013. Retrieved 23 December 2018 from https://www.nparks.gov.sg/florafaunaweb/flora/1/9/1924

Singapore National Parks Board, '*Dendrobium pachyphyllum* (Kuntz) Bakh.f.', NParks Flora & Fauna Web, 2013. Retrieved 22 December 2018 from https://www.nparks.gov.sg/florafaunaweb/flora/5/8/5869

Smith, B.R., 'William John Swainson: who was he?', *The Telegraph*, 8 October 2013. Retrieved from https://www.telegraph.co.uk/technology/google/google-doodle/10363024/Who-was-William-John-Swainson.html

Straits Settlements, *Government Gazette*, 25 (13–25) (Apr. 3–June 26,1891), Singapore: Government.

Teo, C.K.H., *Native Orchids of Peninsular Malaysia*, Singapore: Times Books International, 1985.

Teoh, E.S., *Medicinal Orchids of Asia*, Cham, Switzerland: Springer International, 2016.

Teoh, E.S., *Orchids of Asia*, 3rd edn, Singapore: Marshall Cavendish, 2005.

WCSP, World Checklist of Selected Plant Families, Facilitated by the Royal Botanic Gardens, Kew. http://wcsp.science.kew.org/ Retrieved 2 January 2020.

WWF Malaysia (2001). Study on the Development of Hill Stations, Economic Planning Unit, Prime Minister's Department. Final report, vol. 2. http://repository.wwf.org.my/technical_reports/S/StudyOnTheDevelopmentOfHillStations–FinalReportVol2.pdf Retrieved 27 April 2020.

GLOSSARY

Bilateral	having two sides.
Bilobed/Bilobulate	having two lobes.
Bract	a leaf-like organ subtending a flower or inflorescence.
Callus	a fleshy ornament on the lip.
Carpel	female reproductive organ of a flower.
Caudate	having a tail-like appendage.
Column	a tubular waxy structure formed by the fusion of the male and female parts of the flower.
Concave	a surface that curves inwards.
Convex	a surface that curves outwards.
Deciduous	shrubs or trees shedding leaves during certain times of the year.
Dipterocarp	a family of hardwood timber trees with two-winged seeds distributed through the tropical regions.
Distal	situated away from the point of origin.
Dorsal	the upper side or back of the plant or flower.
Elliptic	oval and flat in a plane, narrowed to each end.
Epithet	the part of a taxonomic name identifying the subordinate unit (species) within a genus.
Epiphyte	a non-parasitic plant growing on another plant (usually a tree).
Falcate	curved or hooked.
Family	a taxonomic category that ranks between order and genus. A family may be divided into subfamilies.
Genus/Genera	a category of biological classification ranking between the family and the species.
Glabrous	having a surface without hairs or projections.
Globular	having the shape of a globe or a spherical form.
Gregarious	in reference to flowering, a mass flowering event where all plants of a particular species bloom at the same time.
Holomycotroph	a plant that does not synthesise its own food but obtains its nutrients through symbiosis with fungi.
Incurved	curved inwards.
Inflorescence	A cluster of flowers arranged along a single or branching axis.
Internode	an interval or part between two nodes of a stem.
Keel	a lengthwise ridge-like structure.
Lamina	the expanded part of a leaf or a petal.
Lanceolate	a narrow oval shape tapering to a point at each end, like a lance.

Lip/Labellum	modified median petal opposite the fertile male reproductive part.
Lithophyte	a plant growing on a rock.
Lobule	a small lobe.
Median	located in the middle or in an intermediate position.
Mentum	a chin-like extension at the base of the flower formed by a converging column-foot, lip and lateral sepals.
Midrib	a large strengthened vein along the midline.
Monophyletic	a group of organisms consisting of all the descendants of a common ancestor.
Nectary	a nectar-secreting glandular organ in a flower.
Oblanceolate	inversely lanceolate.
Oblique	having the axis not perpendicular to the base.
Oblong	an elongated circle.
Obovate	inversely ovate.
Obtuse	a blunt end, between 90–180 degrees.
Oval	a closed curved plane that outlines the shape of an egg.
Ovate	a broader, egg-like basal end with a short tapering apex.
Ovoid	resembling an egg in shape.
Pedicel	a small stalk bearing an individual flower in an inflorescence.
Peduncle	the stalk of an inflorescence.
Pendulous	hanging downward.
Petals	the modified, often brightly coloured, leaves that surround the reproductive organs of the flower, directly inside the sepals.
Pilose	covered in long, soft hairs.
Plicate	with a series of longitudinal folds or pleats.
Pollen/Pollinia	a fine to a coarse powdery substance comprised of male gametes/ a coherent mass of pollen grains produced by each anther lobe of the male reproductive part.
Pseudobulb	the thickened base of a stem which acts as a storage organ.
Pseudoraceme	presence of multiple flowers at each bract axil that are spaced out equally.
Racemose	having the form of a raceme, i.e. a simple inflorescence in which the flowers are borne on short stalks of about equal length at equal distances along an elongated axis and open in succession toward the apex.
Recurved	curved backwards.

Resupinate	the characteristic where the entire flower is inverted with the lip positioned lowermost. This happens because the pedicel of the flower is twisted during development.
Revolute	curved or curled back.
Rhizome	a modified, buried (or partially buried) plant stem that sends out roots and shoots from its nodes.
Saccate	having the form of a pouch or a sac.
Scape	a long internode that forms the basal part or the whole of the peduncle.
Self-pollination	the pollination of a flower by pollen from the same flower or from another flower on the same plant.
Sepal	outermost part of a flower that protects the flower bud, directly outside the petals.
Spathulate	shaped like a spatula; an extension that becomes broader and rounder at the apex.
Species	a group of living organisms consisting of similar individuals capable of exchanging genes or interbreeding.
Spur	an elongated saccate extension at the base of the lip, often containing nectar.
Stamen	pollen-producing male reproductive part of the flower.
Staminode	protuberance on each side of the fertile anther formed from non-functional anthers.
Stelid	pointed projection arising at the column apex.
Stigma	the part of the flower that receives the pollen during pollination.
Subfamily	a taxonomic category that ranks below family and above tribe or genus.
Synsepalum	a structure formed by the fusion of two or more sepals.
Terminal	growing at the end of a branch, stem or inflorescence.
Terrestrial	a plant growing on the ground, usually with its roots in the soil.
Tetrad	a group of four cells produced by a successive divisions of a mother cell.
Trilobe	having three lobes.
Tuber	a short, fleshy underground stem bearing buds which have the potential to produce new plants.
Umbel	a racemose inflorescence in which the pedicels arise from the same point to form a flat or rounded flower cluster.
Ventral	located on the lower surface.

INDEX

REXY PRAKASH CHACKO

Rexy Prakash Chacko is a nature-loving electrical engineer. He is the co-founder of Penang Hills Watch, a founding member of the Penang River Awareness Project and an active participant in Penang's vibrant civil society. He developed a fascination for orchids at the age of 8 when he was gifted a few plants. His passion for hiking the hills of Penang opened his eyes to the diverse species of wild orchids that exist in the hills. He is also the author of *Nature Trails of Seberang Perai* (2019), co-author of *Peaks and Parks* (2019) and *Creating Future-Proof Cities: How to Navigate the Climate Crisis* (2019).

SANTHI VELAYUTHAM

Santhi Velayutham is the Assistant Curator of The Habitat Penang Hill. A trained botanist, she fell in love with plants at a young age and went on to graduate with a bachelor's degree in Plant Resource Science and Management from University Malaysia Sarawak (UNIMAS). She was previously attached to the Mushroom Research Centre at the University of Malaya (UM) as a postgraduate research assistant where she collected, identified and hybridized mushrooms. Her keen interest in native orchids began when she became Assistant Curator at The Habitat Penang Hill and stumbled upon the treasure trove of orchids in the park. Santhi is also passionate about trees and is an International Society of Arboriculture (ISA) certified Arborist.

Entrepot Publishing Sdn Bhd was established in Penang, Malaysia, in early 2015. Its core business is the distribution and publication of fine books on a variety of subjects, with a concentration on Southeast Asia. Entrepot's founders, being authors themselves, have a deep understanding of the essential ingredients that result in fine books: thorough research, quality writing, editorial excellence, superb graphics and world-class printing. Entrepot Publishing is also keen to nurture local talent in such fields. Our ethos is simple – quality service and a quality product.